The Slow Plague

B

Liber geographicus
pro bono publico

The Slow Plague

A Geography of the AIDS Pandemic

Peter Gould

First published 1993

Blackwell Publishers
238 Main Street, Suite 501
Cambridge, Massachusetts 02142
USA

108 Cowley Road
Oxford OX4 1JF
UK

Library of Congress Cataloging-in-Publication Data
Gould, Peter, 1932–
The slow plague : a geography of the AIDS pandemic / Peter Gould.
p. cm.
Includes index.
ISBN 1-55786-418-7.—ISBN 1-55786-419-5 (pbk.)
1. AIDS (Disease)—Epidemiology. 2. Medical geography.
I. Title.
RA644.A25G685 1993
6147.5'993—dc20

92-38653
CIP

British Library Cataloguing in Publication Data
A CIP catalogue record for this book is available from the British Library.

Typeset in 10 on 12pt Sabon by Acorn Bookwork, Salisbury
Printed in the USA

This book is printed on acid-free paper

For
Rod and Deb Wallace
who somehow find
the courage
to keep going

Contents

List of maps and figures

Preface: why a geographer writes about AIDS

Over the past few years a number of people, both in and out of the academic world, have expressed everything from simple curiosity to outright puzzlement about a geographer investigating and writing about the AIDS pandemic. Usually their perplexity arises because they do not have a very clear idea about what geographers do in general, or the way geography provides a special perspective on our world. Few would have much problem with an historian writing a history of the pandemic, even if that history were quite recent, perhaps still immediate and on-going. Most people are used to narratives through time and take them for granted. After all, telling careful stories about human life over time is what historians do. Everyone knows that.

When you point out that, yes, things happen over time, and so always have a history, but that they also happen in space at particular places, so they must always have a geography, people either do a "Huh?," or sit back and say "Hm, yes, I never thought about it like that before." That lots of people have not thought about it like that before is not their fault. In many countries, geographic education is poor, or boring, or both. It is reduced to litanies of disconnected facts, rather like teaching the marvelously complex and colorful panorama of history as lists of dates to be memorized. I suggest that we have all suffered from such pedagogues at some point in our lives, and that we agree that they do not deserve the honorific of "teacher."

But there is another, and still rather puzzling reason for the lack of geographic perspective. Something happened in the early nineteenth century as the human sciences and humanities began to emerge from the flux of philosophical inquiry and into their own separate identities in the universities – economics, psychology, anthropology, sociology, political science, and, surprisingly, even geography and history. I say "surprisingly" because formal geographic and historical studies were pretty well recognized by the end of the eighteenth century, even if you did not find formal departments of these in the universities. But as these disciplines crystallized out of the emerging flux, disciplines that we simply take for granted today, something happened to split geography and history apart – not to say anthropology, economics, and the rest. In the process we seem to have lost sight of the importance of space and geography amid a more general concern with time and historical development. The result, very much to our intellectual detriment, is that we have tended to become a world that is intensely aware temporally, but still almost oblivious to the spatial dimensions of our lives. Why this happened, why spatial awareness was severed from the temporal, is still something to be carefully explored and explicated by intellectual historians, most of whom, fortunately, refuse to acknowledge the academic divisions of knowledge that increasingly have hardened into mere administrative conveniences or disruptive academic fiefdoms.

Fortunately this situation is changing rapidly today, and along all sorts of dimensions. International comparisons point up glaring deficiencies in geographic awareness in children, many of whom even lack what geographers somewhat disparagingly call "capes and bays" geography, some bare minimum knowledge of places that helps people give meaning to reports about their world. The various environmental crises have also generated a much richer sense of global and geographic awareness, as such destructive changes as pollution, deforestation, ozone holes, nuclear disasters, and global warming, make more and more people realize that what happens at one place on our planetary home can affect many others far away.

But I think a third major reason for the heightened awareness of geography and the geographic perspective is that the new generation of geographers is much less parochial intellectually, and takes genuine delight in reaching out to others. Fifty years ago, many prominent geographers were terribly defensive, spending much of their time defining where geography began and where it ended. A lot of professional geographic writing was a sort of intellectual urination around the territory to mark what was "real geography," indicative of a defensive us-versus-them complex. All that has changed, both to the benefit of geographical inquiry, and to adjacent areas receiving an infusion of "spatiality" to augment their "temporality."

Now the last thing I want to suggest is that geography and geographers deal exclusively with space and place and ignore other perspectives, or that these concerns are somehow exclusively theirs and theirs alone. That would be just as silly, contradictory and dysfunctional as an earlier generation's boundary-marking. As human beings we exist in space and time, and we also have the ability to think. In some simple, but quite profound sense, we are all geographers and historians and philosophers, and have the capacity to sharpen our thinking and understanding along all sorts of lines and directions of inquiry. Perhaps the only thing that marks professional geographers from the rest is an absorbed fascination with that spatial domain, though certainly not to the exclusion of any other perspective that will help us light up this wonderful and complex world in which we find ourselves.

So despite strictures, warnings and disparagements from tribal elders, several research traditions in geography have adamantly refused to sever time from space, the history of places from the larger spaces in which those places are embedded. Historical geographers, for example, explicitly incorporate history and time into their very title, but others have also been fascinated with tracing human developments through time and over geographic space. If it did not sound so jargony, I would really like to call these geographers "spatio-temporalists," but we will put that term aside for later, and simply note that most of them seem to be associated with inquiries that concern the *diffusion* of things. These may be anything from innovations, ideas, new crops, new technologies to . . . and – here it comes – new diseases. But it is the *process* of diffusion, rather than its content that is of particular interest. Many of these studies are descriptive, teasing out on maps, in written text, and in algebraic equations (often all three), the way things spread. Some are highly formal, writing out mathematical "rules of the game," and then testing these with computer simulations. Others actually twist, reshape and transform the familiar map to help the geographer account, in a simplified but more understandable form, for why things happened to occur as they did at particular times and at particular places. In passing, it is worth noting that the first computer simulations in the entire social and behavioral sciences were undertaken by diffusionist geographers, on machines that today are truly museum pieces.

Given this long and rich tradition, it should really come as no surprise that a geographer writes about the AIDS pandemic, or that the geographic perspective focuses a rather different light on it compared to traditional (and they can be very traditional!) people in medicine and epidemiology. Many of the things I am writing about are based on personal experience and highly specialized research. But I have a very strong feeling that from time to time people in universities should climb down that circular staircase in the ivory tower, and try to reach out beyond its academic walls to let other people know what has been seen from its vantage point. Perhaps it is worth reflecting upon the fact that "specula," the root of "speculation" is a *watchtower*. That is why this book is one of a series labeled *liber geographicus pro bono publico* – a geographical book for the public good, which sounds just a bit pretentious until we translate it more loosely as "a book for the busy but still curious public." I cannot in all honesty say "I hope you enjoy it." It is a book about a catastrophic pandemic, about life and death, of great courage and not a little perfidy. But I do hope it lights up one aspect of this terrible disease in a way that often seems to have been suppressed before.

Acknowledgements: intellectual antennae

As a geographer, I have the great privilege of teaching and learning in a large university, and of belonging to some of those invisible colleges whose threads of shared concern and delight encircle the globe. Inevitably, this means that as soon as you start to work on something others soon hear about it, and in their generosity they become your intellectual antennae somewhere "out there," picking up pieces of information, new research findings, and conflicting perspectives that otherwise you might have missed. A world pandemic today produces millions of words each week, and even if many of them are repetitive and redundant, no one can possibly keep up with all of them. This means that any account is selective, partial, and written from a particular perspective. It cannot be otherwise, but it helps to have friends and colleagues sensitive to your interests. I have lost count of the references, articles, reports, and many other sources of information that came in an envelope, during a telephone call, stuck in my box, shoved under my door, or in conversations. I cannot possibly acknowledge them properly, but to all that helped in these ways – please accept my thanks.

Nevertheless there are just over a baker's dozen whose contributions to this account must be named – or there is no justice in this world. At Penn State, Deb Straussfogel was central to some of the earlier experiments in computer modeling, and she will recognize her efforts in chapter 6. Joe Kabel has been my right-hand man over the past four years of modeling and research, teaching me far more than I ever taught him – always the sign of a good graduate program. He will see his efforts in chapter 14. More recently, he was joined by Ralph Heidl and Bill Holliday, whose shared concern for the graphic visualization of important problems was augmented by computer skills capable of handling large amounts of geographic information. They will recognize their expertise underpinning a number of the maps in this book. Equally aware that geography matters, Terry Dawson recognized how the issue of confidentiality was totally embedded in this way of looking at the world, and she will see herself reflected in chapter 13. Beyond my own university, I want to acknowledge the privilege of working with Wil Gorr at Carnegie-Mellon University, the person who pioneered the most promising way we presently have of forecasting what maps of epidemics are going to look like some years into the future. We tried hard to engage the interest and purse strings of certain powers, and although our joint efforts did not bear fruit I shall always cherish my association with a generous and open scholar, and a fine teacher. Woody Pitts deluged me with references from Hawaii and California until my bibliographic cup really did run over, while Bill Bowen at Northridge generously shared both his time and his remarkably detailed cartographic insights. John Thompson scavenged the local reports for me wherever his travels took him in North America, Europe and Africa, and Rich Symanski has needled me with a typically "rich" correspondence that sometimes made me think again.

Singling out one person might seem insidious, but it is not. Kerry Demusz served as an honorary research assistant in a faraway land, and quickly became a collaborator and thoughtful critic. I shall never know how she compiled her monthly reports for me in the midst of learning her second, non-Western language, all the while plunging into AIDS-related activities involving the empowerment of women. She is the stuff senators are made of. I hope you vote for her one day.

Lori Lynch did all the typing and revisions with infinite patience and good humor, and she has my heartfelt thanks.

Finally, two others will always represent for me the very best of socially-engaged scholarship confronting Establishment mendacity and bureaucratic apathy. I hope the dedication of this book indicates the respect that this spoilt brat in his ivory tower has for those in the front lines trying to create a truly civil society.

Hatteras, May 1992

Prologue: new plagues for old – the horseman rides again

Ring-a-ring-a-roses,
A pocket full of posies.
Atishoo, atishoo
We all fall down.

Children's song from the plague years

In memory – a collective memory impressed upon materials as varied as papyrus, vellum, paper and magnetic disc – the human family recalls many plagues. Like all memories, these traces of remembering fade with time, becoming a whisper, even silence, in preliterate cultures after a few generations, or maintaining themselves in faint, ghost-like forms for millennia in those with the power to support memory with pictures, characters and letters. Some of the oldest writing records the trace of scourging pestilence, and the memory sharpens, becomes more detailed and precise, as we approach our own printed times.

Only 700 years ago plagues moved slowly, inching across the land with the plodding caravans, or with the speed of other carriers such as rats and insects. Three forms of bubonic plague, known collectively as the Black Death, took about 12 years to move over land from China, until they reached Asia Minor and the Black Sea in 1346, infecting the sailors and merchants of the trading posts. Once on water the Black Death moved faster: the next year it was in Sicilian and other Mediterranean ports, poised for a great flanking attack by sea along the coastline of Europe, ready to sweep northwards across the continent, pushing in front of it a great wave of well-founded fear. By 1348, coastal villages as far away as England were infected: those of Norway and Sweden followed in successive years.

Then, in 1493, a new plague arrived: the "pox" appeared in the ports of Spain, carried by the sailors of Columbus from a new and unsuspected world. By 1495, it had become the "Naples Disease", carried back to France by the soldiers of Charles VIII after their campaign. Only 25 years later, it was reported by the Portuguese in Goa and Macau, half a world away. We have lived with syphilis ever since. And died with it too. In the 1660s another of many waves of bubonic plague came to Euope, this time the Great Death, the carts piled high once again with the dead, stopping at more and more doors marked with the X of death rather than the cross of life. And so the story continues to our own times. Typhoid, typhus, smallpox, measles, cholera and influenza . . . each disease in its turn sweeps across the map, moving through time and over space to give every epidemic a history and a geography.

Ours is an epoch when both historical time and geographical space become more and more structured and conditioned by human technology. Distance, that old separating protective distance, begins to contract as the technology of sailing ship, hardened road, steel rail, automobile and airplane shrinks the globe, tightening the physical bonds connecting the human family. Space is compressed, and time speeds up. By 1866, Adrien Proust, father of Marcel, knows you can only stop a plague by severing yourself from the connecting structures that support and carry the traffic of disease. The farther away you can break the chain the better. The *cordon sanitaire* is his idea, as he declares that Egypt is the first line of defense for Europe against new epidemics like cholera. His own extensive travels in Persia try to elucidate the old pathways, north to Russia, and west to the Mediterranean, pathways of infection for Europe and North Africa over the centuries.

Many of these plagues are still with us, though contained by public health measures, vaccines and cures that are simply taken for granted – until, in 1918–19, an influenza pandemic kills over 30 million people, or cholera sweeps though Latin America in 1991, carried by contaminated streams in Peru to the huge Amazon basin, or by infected air travelers spreading it from city to city. Fortunately, one plague has disappeared forever: smallpox has gone, squeezed relentlessly into a smaller and smaller area in Somalia until in 1977 the last potential carrier was vaccinated, and smallpox finally had nowhere else to go. Measles, another disease surviving only in a human host, is next on the WHO's (World Health Organization's) list to be eliminated. For those diseases capable of surviving outside of the human body – cholera in water; rabies, malaria, bilharzia, and many more in other living hosts – the problem of eliminating them is much more difficult, perhaps impossible.

Of all diseases, those caused by the viruses are often the most difficult to eradicate, not the least because many display a distressing capacity to mutate, to change their properties, so that no sooner do we develop a vaccine against a particular type than another one appears immune to our efforts. Influenza is a good example, as each year we try to stay one step ahead of it by producing vaccines to counter the latest strains reported. But of all the diseases caused by viruses, the most insidious are those with long incubation times. With most diseases, the common symptoms usually appear quite quickly in someone who has been infected – two or three hours, two or three days, two or three weeks at most. You know, and usually the people around you know, when you come down with one of these. You feel ill, take to your bed, and generally disconnect yourself from most of society, taking yourself out of circulation until you get better – or die. Quarantine, either self-imposed, medically recommended, or mandatory under law, essentially disconnects the infected person from those still healthy.

But the viruses with long incubation times pose much more difficult problems. After transmission and initial infection they may lie dormant, producing no distinctive symptoms, or remaining otherwise undetectable even to sensitive tests. There they lie, sometimes for years, waiting for a trigger to set them off. In the meantime, those infected can transmit them to others, and so on in turn. In 1981, just such a virus was suspected in a number of cases of pneumonia and cancer contracted by young men around Los Angeles. These were such rare forms that they were seldom seen by doctors in a lifetime of practice since they were usually held at bay by the body's natural immune system. By 1982, mainly because of work in Japan, France, Britain and the United States on related viruses, the first type of human immunodeficiency virus (HIV-1) had been identified. Three years later, in 1985, French virologists identified a second – HIV-2.

The human immunodeficiency viruses are the cause of the new plague, a plague whose final toll already threatens to dwarf even the 30 million deaths of the influenza pandemic of 1918–19. It is a slow plague, but a sure plague. The average time between initial infection and the collapse of the immune system is about ten years: death from once-rare, but now symptomatic pneumonias, cancers and brain infections almost invariably follows within three or four years – or only six months if you are a small child. Despite the media focusing on a small number of increasingly rare cases, few people survive much longer. We are dealing with a mortal disease. On this small planet, where hardly anyone is more than a day away from anyone else, we have a new plague.

1

The killer: HIV and what it does

I do nothing upon myself, and yet am my own executioner.
John Donne, Sermons, *1640*

When you put money, disease and human interests together, you get an explosive situation.
Simon Wain-Hopson, Pasteur Institute, Paris

This is not a mystery story, we know who the killers are. The HIV-1 and HIV-2 were identified in 1982 and 1985, thanks to brilliant detective work in virology spread over three continents. Unfortunately, the story of research is deeply blemished by overwhelming arrogance, false claims, catastrophically dysfunctional rivalries, false published evidence, greed for scientific recognition, avaricious claims for huge amounts of money, and governmental interference that had everything to do with national convenience and nothing to do with the truth. It forms a superb, if extreme, example of the fact that science is always a socially negotiated and socially interpreted endeavor.

And there we will leave it, with all its potential for satisfying prurient curiosity, waiting for the day when historians of science unravel what can only be described as an unsavory mess. It has its heroes and villains, those who behaved well, and those who did not. It has nothing to do directly with the story here, but we should be aware that sometimes, underneath the cool descriptions of scientific research and its brilliant discoveries, there are intense human dramas.

Finding the killers was no easy task, and explicating them, in the literal sense of unfolding and describing the viruses and the way they act on the human body, raises questions of great difficulty. In many areas of science today research has become so advanced, and knowledge so specialized, that even those who appear to be in closely adjacent fields find it difficult to communicate their findings to each other.

Some years ago I attended a conference whose purposes were to lay out guidelines for national policy on the AIDS epidemic based on the current state of the various sciences involved. We were to indicate the possible directions research might take, and then estimate the funds needed to sustain such huge programs of inquiry. Those at the conference were a mixture of people in medicine, epidemiology, virology, mathematics, biology, sociology, and so on, all scientifically trained and highly competent in their respective fields. Each day, after meeting in six specialized groups, we met together to inform each other about what was going on, what the state of the art was in our particular area of expertise. One virologist, working on what were then the frontiers of research on the HIV, tried to tell the rest of us what had been happening. He spoke well, it was clear he was trying to simplify very complex things for the non-specialist, and one had an intuitive sense that each point he was making must somehow follow logically from the previous one. His account left most of his audience, including many of those in medicine and biology, absolutely baffled. In the murmur of conversation after the session you could hear, "Did *you* understand?" and "What the hell was the guy saying?"

But tens of millions of people around the world are going to die from this slow plague, and we owe it to ourselves, and to our stricken fellow human beings, to get some understanding of the killer and how it works. After all, and however you want to phrase it, there but for the grace of God, or the luck of the draw, go you and I.

The human immunodeficiency virus is very beautiful, shaped like a tiny ball made up of an outer covering of protein molecules forming pentagons and hexagons, and containing within it other layers, rather like those intricate balls-within-balls the Chinese used to carve from a single piece of ivory. Its diameter is about 100 nanometers, a measure so unfamiliar and so far removed from everyday life that it has no meaning for most of us. Saying it is one billionth of a meter does not help much either, but if we made a square box one thousand HIV diameters on each side, and crammed a billion viruses into it, we might just be able to see it as a dot about a quarter of the size of the period at the end of this sentence. How people actually discovered the form and function of something so small, intricate and deadly is one of the marvels of modern virology.

To simplify something very complex, we might say that the HIV, like any virus, is nothing more (but equally nothing less!) than some genetic material – just two strings of genes – wrapped in several coats of fat and protein (figure 1.1). Its deadly nature lies in those strings of genes, together with an enzyme called reverse transcriptase. Until the late 1960s, virologists and geneticists thought that all genetic information was contained in the now-famous double helix of DNA, and that the information was carried to the protein molecules of a cell via an intermediate form called RNA. It was then discovered that a class of viruses existed that only had RNA, the "intermediate" form, as their genetic material, but the accompanying enzyme allowed the RNA to copy itself as a single strand of DNA, and then make another to form the familiar "double helix." The usual genetic information flow, DNA to RNA, was reversed – hence "reverse transcriptase," something that writes and transcribes backwards – and the viruses became known as retroviruses.

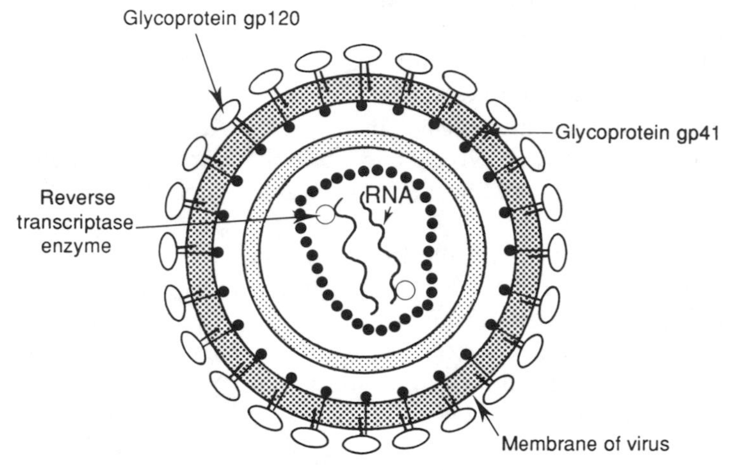

Figure 1.1 The basic structure of the human immunodeficiency virus (HIV), two strings of genes enclosed by a shell of fat and protein, with sugar protein molecular "keys" sticking out from the surface.

Now specialists in cancer research already knew that some sorts of viruses could cause tumors and leukemias in animals. In fact behind this knowledge lies a tragic story in the history of science. As early as 1910, Peyton Rous demonstrated that cancerous tumors in chickens could be caused by a virus (a retrovirus as it turned out much later), but his claim was met with such condescension and disbelief from the establishment that he gave up his research. Like Mendeleev in Russia, whose chemical table was so ridiculed by the chemists of his day that he nearly committed suicide, Rous's demonstration did not fit into the accepted set of beliefs, the "paradigm" as it would be called today, and his remarkable discovery was ignored for more than 40 years. We shall see later (chapter 12), that clinging onto outworn paradigms is still characteristic of some research on the AIDS epidemic. We know now that retroviruses cause cancers in many species of animals, and also in human beings – ever since a retrovirus causing leukemia was discovered in 1978. It took only four more years before another human retrovirus came to light. It was the HIV-1, and here is how it does its terrible work.

Sticking out of the first layer or envelope of the virus are a series of small knobs each made up of a simple sugar protein called glycoprotein, or gp120. They are attached to the outer membrane of the virus by another protein, gp41, that actually penetrates the outer layer. The protein gp120 has a particular propensity to attach itself to molecules called CD4 which lie on the surface of a class of white blood cells called the T4 lymphocytes (figure 1.2). It is rather like a specially shaped protein key trying many molecular locks at random until it finally clicks into place. The T4 cells are called "helper cells" and they play a crucial part in the highly complex human immune system. The HIV ultimately destroys the T4 cells, the very cells that normally help the body to trigger a response to all sorts of infections. Nor does it stop there: any cells in the body with CD4 receptor molecules can be invaded by the HIV, including two others in the first line of the immune system's defense, the monocytes and macrophages. The former roam around the body in the bloodstream searching for infective agents to engulf and destroy, while the latter locate in specific places and destroy infections specific to particular organs. Entry of the HIV into the monocytes and macrophages is particularly insidious because they are not easily killed, and they can harbor the HIV for a long time, allowing it to reproduce inside them and protecting the HIV from any immune response.

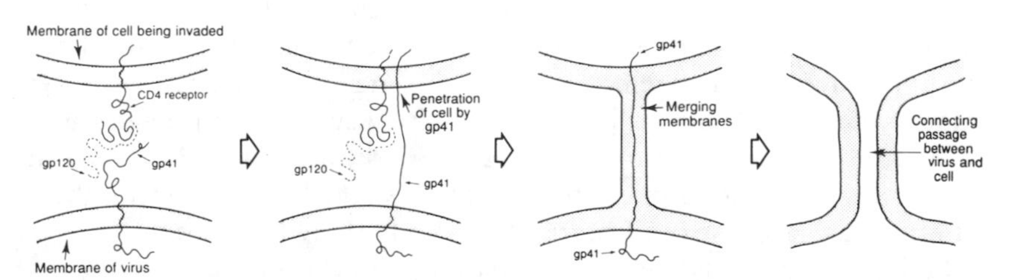

Figure 1.2 How the HIV, with its sugar (glyco) proteins gp120 and gp41 penetrates a cell to provide a passage for the RNA and accompanying reverse transcriptase.

When one of the gp120 "keys" protruding from the surface of the HIV locks onto a CD4 receptor of a lymphocyte, the anchoring root of gp41 follows and penetrates the membrane of the helper cell (figure 1.2). It is along this line of penetration that the two membranes – the virus's and the cell's – merge to form a connecting structure or passageway for the RNA, and its accompanying reverse transcriptase, to move from the virus to the cell. Once inside all hell breaks loose: the reverse transcriptase "writes back" the RNA into a double helix of DNA, and it is this that works its way into the DNA of the cell's nucleus. And there it can sit for weeks, months, or even years, a tiny template for a virus ticking away like a time bomb until the lymphocyte is called upon to fight some infection. It seems to be this that triggers activity again: the DNA starts to produce strands of RNA, each of which can become the core of a new HIV. Other genetic information builds up the layers around it, until all it lacks is its own outer protective membrane. To manufacture this it uses chemicals in the membrane of its host cell as building materials, forming a small protrusion or bud on the cell's surface. Finally, as a new and complete virus, it bursts through, ready to find and lock onto a CD4 molecule on the surface of another T4 helper cell. And so on . . .

The effect of this reproductive process is devastating for the helper cells. They cannot perform their basic function of helping other lymphocytes to produce antibodies to a particular disease (hence their name "helper cells"), nor can they produce interleukin, an important substance that stimulates the immune system to fight infection. And as the HIV reproduces inside them the cells also start to produce proteins on their surface that bind to other, still uninfected cells, until great clumps of these, called syncytia, form. Most of these conglomerations of cells die quite quickly, as do individual T4 cells, because their membranes are constantly ruptured by HIV buds, each one of which leaves behind a tiny lesion. It does not take many of these before a cell collapses. As for the monocytes and macrophages, they are more resistant, but they can carry many HIVs inside them to all parts of the body, including the brain and nervous system. In the brain, the microglial cells become infected, producing severe brain and other neurological dysfunctions. AIDS dementia is now a definitive symptom in diagnosis.

And so the HIV continues its deadly work, attacking the very cells that stimulate the immune system, and using others that directly fight an infection as carriers to all parts of the body. Slowly, as the T4 helper cells decline, the body's immune response to infections normally kept at bay becomes weaker and weaker. Pneumonias and cancers that were once rare in the medical literature have now become a way of identifying people with AIDS. Kaposi's sarcoma, for example, was never seen by most doctors outside a broad belt across tropical Africa, and the few cases that did appear in the early medical literature were generally men with a Jewish and East European background, leading to the idea that there might be some sort of genetic propensity for it. The blotches on the skin, the "AIDS cancer," are now found in 15 percent of the cases under 60 years old. Pneumonias, particularly a form *pneumocystis carinii* that is difficult to treat, affect over half the cases. In fact it was the sudden appearance of five cases of *pneumocystis carinii* in Los Angeles in late 1980, and the urgent call for a then-experimental drug dispensed only by the Centers for Disease Control (CDC), that alerted the medical world to the possibility that a new disease had made its appearance. These, and many other symptoms of once rare but opportunistic infections appear to develop more quickly in people with AIDS who are already infected with other diseases, for example hepatitis B and some of the herpes viruses. It is these that trigger the production of new HIV in the T4 helper cells, producing a much quicker build-up in the body, and a more rapid collapse of the immune system.

As a retrovirus, the HIV belongs to a large group called the lentiviruses, the slow (*lentus*) viruses that have the ability to enter cells without initially doing too much damage to them, and then lying there for long periods of time before starting their final and explosive burst of replication. Lentiviruses of different sorts affect many animals – goats, sheep, horses, cats, monkeys, etc. – often with similar symptoms to AIDS in human beings – pneumonias, lymph gland infections, wasting, cancers, and neurological disorders. All are symptomatic of the immune system going down and being unable to cope with infections, and all appear to have been around for a long time. But HIV appears to be a new virus, the cause of a new plague. When did it arrive? And where did it come from?

2

The origins of HIV: closing an open question?

So geographers, in Afric's maps,
With savage-pictures fill their gaps;
Jonathan Swift, On Poetry

The retroviral map is like a map of the world in the fourteenth century. Some coastline is well-defined, other portions are blank and infested with dragons.
Jonathan Mann, before resigning as Director of the WHO AIDS *Program*

Scientists are trained to be conservative and patient. If you do not know, if you are not sure, you try to find out by marshaling evidence so convincing that even the most doubting Thomas among your critics has to concede the truth of the story you tell. Now that phrase, "the story you tell," will not please many scientists who have received a highly specialized, and by definition rather narrow, education. Scientists, so the mythology goes, do not "tell stories." Stories are things you tell children at bedtime. Scientists are seekers after and tellers of . . . the Truth. Their initial speculations, usually rephrased in a grave voice as *hypotheses*, are tested again and again until the truth of God or nature, depending upon one's theological predilection, is revealed. Few would acknowledge that their scientific papers, reviewed and scrutinized by their peers, were simply persuasive fairy tales, mere acts of rhetoric. No matter how honorable the art of rhetoric was in Roman times, "rhetoric" today tends to have a rather theatrical, even dishonest, ring to it. It is used by the politician and seller of snake oil, and sometimes there is not much difference between these two charlatans. The scientist is quite different.

The trouble is that the truthful stories that scientists try to tell at one time often turn out to be wrong at another. Even Newton's story, the one he told in his *Principia* in 1689, was eventually shown to be quite wrong by Einstein. And you cannot just make a few corrections here and there, dust Newton off, and set his theory on its feet again. Those feet are made of clay: Newton simply did not get it right, although his rhetorical story was so convincing that it became the paradigm, the only framework within which questions could be asked, for over 200 years. It is the same in medicine. One dean of a medical school, in a surprising burst of honesty that tends to be rare in a profession not noted for its modesty, told a graduating class of medical students that half the things they had been taught would be proved wrong in about 20 years time. "The only trouble is," he said, "we don't know which half."

Scientific questions, as open speculative questions, are difficult to close. So it is with the question of the origins of HIV. It may well be that we can never hope for *the* definitive answer, but this does not mean we have to throw up our hands in despair. Pieces of evidence accumulate and begin to fit together like a jigsaw puzzle, and the emerging picture becomes clearer and clearer, more convincing and meaningful, even if there are pieces still missing. Of course, we may find out in the future that we have been forcing bits from different puzzles together, that some pieces are lost forever, and that we shall have to revise our interpretation of the evidence and start telling another story. It happens to the best, as many bold and magnificent failures in science will tell you. But with lots of bits of evidence around, and more and more coming to light each year that seem to fit into place, we do not have to sit around tongue-tied, even if the story may anger or embarrass some because they find it politically unacceptable.

Let me say it clearly: the origin of the HIV, both the 1 and 2 varieties, was Africa. All the fingers of evidence point to African origins, and they are scientific fingers not accusing fingers. It is important to make this "pointing" absolutely clear, because misinterpretation has had some deadly consequences. The origins of a disease are important to know for two major reasons. First, we want to understand how and where it started, and how and where it spread. These are questions asked by epidemiologists, people who try to generate knowledge (*logos*) about the way a disease moves through (*epi*) a people (*demos*), although most of them remember vaguely only the *when* questions of the historian, and have totally forgotten the *where* questions of the geographer, as we shall see.

Second, if a new virus appears it must be the result of a mutation of an existing one. Alternative possibilities include its creation by God to punish sinful people, nature generating spontaneously a way of reducing population pressures, or its sudden appearance from outer space. All of which have been suggested, but I see no reason to waste time on them. To delimit the region from which a new virus appears is important on scientific and humane grounds because it is here that we are most likely to find the closely related viruses of which the new one is a mutant. The more of these viruses we can identify, and the more we can learn about them, the greater our chances are of developing a vaccine against HIV, or at least something that will block its devastating effect on the human immune system.

No one should blame someone else for the arrival of a new disease. Current revisions of history not withstanding, no rational person can *blame* Columbus or his sailors for bringing syphilis to Europe, or measles to America. There was little understanding then of how diseases spread, and a germ theory was still hundreds of years in the future. Diseases appear: it is nobody's fault. The great pandemic of influenza of 1918–19 was first reported in the beginning of March at Camp Funston, Kansas, spread rapidly to other Army posts, and had jumped to the ports of western France by April on the troopships. No one blamed the hog farmers of America's Midwest for the new and lethal virus, even though it may have mutated in the large pig populations, and no one held the Army guilty of biological warfare. This last point is not entirely facetious: suggestions have been made that the CIA genetically engineered the HIV and released it. But the earliest serological evidence we have for HIV is 1959, when no genetic code was known. Reverse transcriptase was only discovered in 1970, and no retrovirus affecting humans came to light until 1978. Whatever one's views of the CIA, I see no reason to waste time on such an off-the-wall speculation.

Nailing down the first appearance of HIV is difficult. A search of the medical literature back to 1950 raised the possibility of single cases in the United States, Canada and Britain during the 1950s, but in each case the evidence consisted of clinical symptoms of rare pneumonias long reported in the medical literature, with no accompanying Kaposi's sarcoma. The first identification from blood serum, using tests only developed in the early 1980s, was from a sample frozen and stored in 1959 in Zaïre. It was also in Zaïre that a young Danish surgeon was serving in a remote rural hospital, and being unable to diagnose her increasingly severe illness she returned to Copenhagen in 1977. Her immune system continued to collapse, although no one then knew why, and she eventually died. Fortunately, Danish doctors kept meticulous records, and these now reveal all the classic symptoms and progressions of AIDS. It is assumed, quite reasonably, that she was infected by a surgical "stick," blood from a patient infected with HIV-1 entering her body through a cut while operating, in much the same way that the virus travels on the hypodermic needle of a drug addict sharing with others. It is by far the most effective way to transmit HIV. Surgical sticks are not uncommon: surgeons often have to work very quickly in confined spaces with instruments sharper than razors. Gloves are extremely thin to retain as much sensitivity as possible, and they offer no protection against a "stick." Those prepared to operate on people known to be infected with HIV display a cold courage daily that may be beyond most of us.

The evidence for an African origin continues to pile up. The earliest case in Europe was probably an English seaman with a history of African contacts who died in Manchester in 1959. The earliest of which we can be quite certain was a Norwegian seaman who showed symptoms of HIV infection as early as 1966, so he was probably infected in the late 1950s or early 1960s. He traveled to many African ports while a seaman, and had a long history of other sexually transmitted diseases. By 1967 his wife was showing similar symptoms, and by 1969 their two-year-old daughter also became very ill. All three died in 1976 from the now-classical opportunistic infections of AIDS, but the Norwegian doctors, unable to diagnose the reasons for the collapse of their immune systems, had frozen blood serum from samples going back to 1971. All samples from all three members of the family later tested positive for HIV-1. Once again, there appears to be an African connection in the early cases we can positively identify in retrospect with the tests we now have.

A fourth piece of evidence for an African origin is of a very different but thoroughly geographical nature. In the last few decades, particularly with the establishment of the WHO by the UN, and the designation of the CDC in Atlanta as the international center for reporting the outbreak of diseases all over the world, a few epidemiologists have begun to realize that diseases are truly a global phenomenon, that they actually do exist in space as well as time. If we can begin to understand how they spread across the face of the earth, we can use this knowledge as an "early warning system." In 1991, for example, travelers from Latin America to the United States received warning slips about cholera, asking them to report immediately to a hospital or doctor if they experienced symptoms of the disease. In cholera, early treatment may be literally vital, and it is important that an infected person does not contaminate food and drink consumed by others. With other diseases, influenza for example, we may be able to manufacture vaccines in advance of the arrival of a new and virulent strain. Perhaps it is no accident that some of the best work on the spread of influenza was done in the Soviet Union, a huge country transporting many of its people by rail. In the 1970s, Soviet epidemiologists used rail transportation as a basis for a computer model of influenza spreading from the Soviet Far East to the western republics, and in the 1980s, as airplanes became more available for civilians, air transportation was also incorporated.

Computer modeling moved to the global scale in the 1980s, when the structure of the air network connecting 51 of the world's cities was used successfully, by collaborating Soviet and American epidemiologists, to model the spread of virulent, and readily identifiable strains of influenza from Hong Kong. This model of "global connectivity" was then used by French epidemiologists to model the hypothetical diffusion of HIV. In the computer you could start the HIV pandemic at any one of the 51 nodes, and see what the global map looked like at successive times. For most starting points – San Francisco, Casablanca, Havana, etc. – the resulting maps of HIV infection bore no resemblance to even the crudest estimates reported to the WHO. Only when the epidemic was started in Kinshasha, the capital of Zaïre, did the resulting map resemble quite plausibly the known global intensities of HIV infection.

The evidence for an African origin is already overwhelmingly strong, and it is supported even further as we narrow the hunt down, asking where and what were the likely sources of HIV-1 and HIV-2 in Africa. Obvious sources for viruses are always animal populations, for we have long known that the influenza virus can be harbored by pigs, and from them mutate to forms virulent to human beings, while the rabies virus is no respecter of the boundary between wild animal populations – foxes, raccoons, skunks, wolves, etc. – and human beings. When the first retrovirus in humans was discovered in 1978, and named the human T-cell lymphotropic virus (HTLV-1), it turned out to have a surprisingly wide geographic range; from Japan, across the Pacific to the Caribbean and northern Latin America, and then across the Atlantic to Portugal, and, once again, to West and Central Africa. One speculation was that the early Portuguese explorers and slavers had carried it from Africa to other parts of the world, but there is too much awkward and contrary evidence to make this stick. But the Japanese connection proved important for it was in southern Japan that a closely related form of the virus was found in wild macaque monkeys – the simian or STLV – and this discovery raised the question that there might be a closely related simian form of HIV. The HTLV-1 is not an HIV, but as another human retrovirus it does bear a marked family resemblance, both in the way it reproduces itself in the body and causes forms of cancer. Perhaps, so the reasoning went, if the HTLV had a close simian form, an STLV that developed in close relation to it, then maybe the HIV viruses also had close relatives in the wild monkey and ape populations somewhere around the world.

It was a speculation that paid off handsomely in terms of our knowledge. By 1984 a simian immunodeficiency virus (SIV) had been isolated, and after several false starts, caused by wayward transmissions in laboratory animals, it was found in a large number of African green monkeys in the wild. Strangely, it seemed to do them no harm, although it had a devastating effect upon macaque monkeys from Asia. In a similar way, HIV has an ultimately lethal effect on humans, but it seems to be tolerated by chimpanzees who are infected by it in the laboratory. As it turned out, there are a number of forms of SIV, all related in varying degrees, but the form found in another monkey population, the sooty mangabey, provided the strongest geographical clue to the origins of HIV. Not, at this point, HIV-1, the Central and East African variety, but HIV-2.

The sooty mangabey is native to West Africa, the geographical origin of HIV-2. If you test SIV and HIV-2 using standard, and now very precise serological tests, you cannot immediately tell them apart. Only if you go deeper into the virus, and patiently identify the genetic material called the nucleotide sequences at the core, can you tell that they are not quite the same. In fact, and quite paradoxically, even the "same" viruses may differ slightly, just as we are all people, although individually we are all slightly different. If we think of similarities and differences between viruses represented by some sort of "genetic distance," then there are actually greater distances or more variation in a collection of SIVs from sooty mangabeys then between them and HIV-2s isolated from infected people. We can even think of viruses existing in some sort of tiny micro-geographical space (figure 2.1), and make a map showing how similar or close together they are (short genetic distances), or how different and far apart they are (long genetic distances). It is no accident that geneticists use such geographical metaphors as "maps," "distances" and "regions" frequently in their descriptions of this tiny "world." The closeness of SIV and HIV-2 in genetic space points immediately to cross-transmission from the monkey population to human beings. Or was it the other way around? We shall examine this question in a minute.

But what about HIV-1, the East African variety? On our genetic map it seems as far away from the HIV-2 as Senegal and Guinea-Bissau are from Zaïre and Tanzania on the geographic map. As late as 1985, no HIV-2 had been found in East and Central Africa, and little HIV-1 in West Africa, although with air travel today mixing was already underway. For several years the hunt for a closely-related simian form of HIV-1 went on, including contacting and testing virtually isolated groups of pygmies deep in the forests of central Africa. They regularly hunted and ate monkeys, and if any crossover between species had occurred they might have formed a reservoir of HIV-1 in humans

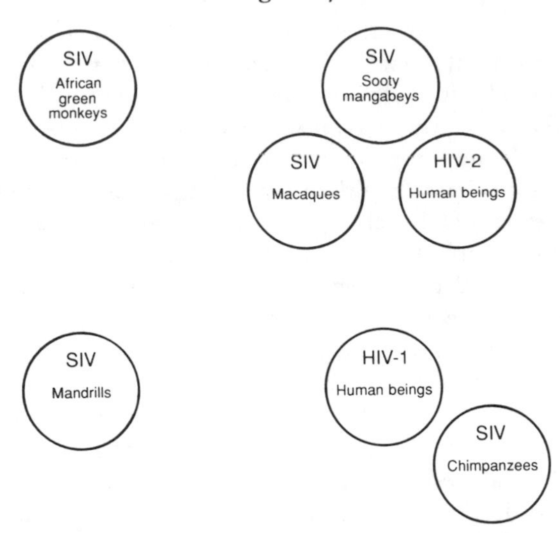

Figure 2.1 A genetic map of some of the immunodeficiency viruses, where similarities and dissimilarities are represented by distances. To show all the similarities and dissimilarities properly, the "map" should probably be in a four-dimensional space. These cause problems for publishers, so we can only show an approximate two-dimensional projection here.

from which a slow and sporadic leakage had taken place to surrounding populations. Four years of testing 782 pygmies brought no HIV-1 to light, although in 1987 one woman came up seropositive. But the transmission proved to be the other way; from an infected non-pygmy man to her, the first, and potentially devastating infection of this one group of people whose bodies and culture had been protected by their geographic isolation, that "disconnection of distance," from others. "No man is an Island entire of itself . . .," wrote the poet John Donne. And no women either. A sexual relation is a connecting relation, and over that connection, forming a sexual structure, the traffic of HIV can be transmitted.

And then, after nearly a decade of searching and testing wild monkey populations to no avail, in 1990 two chimpanzees were captured in Gabon and their blood serums tested for HIV-1. Both samples reacted to the standard test for all the proteins in HIV-1, including the glycoproteins sticking out from the shell or membrane. Closer examination of the nucleotide sequences from a virus isolated from one of the chimpanzees showed that the simian virus was not exactly the human form; in fact, on the genetic map it was a bit farther away from HIV-1 than the sooty mangabey's virus was from the West African HIV-2. Nevertheless it was close, certainly close enough to reinforce strongly the same cross-transmission idea that had been confirmed for HIV-2. We already know, from the proliferation of different SIVs in wild monkey populations, and from the tragic appearance of new varieties of HIV-1 and their acquired resistance to drugs like AZT, that these retroviruses can mutate very quickly. It is possible that they can change three to five times as quickly as the influenza viruses, and these already test our abilities to stay ahead of them. There may not be complete certainty, which is a privilege of the gods not mortals, but it appears fairly sure now that both varieties of HIV were a result of cross-transmission from animal populations in Africa that are, on a much larger genetic map, very close to us.

As for the exact mode of transmission from the monkeys and apes of Africa to us, that is something we shall never answer except with a number of plausible speculations. Wild animals, including monkeys and apes, are an important food source for many African and other people, and in preparing the meat it seems likely that an open sore, or a cut like a surgical stick was the route for cross-species transmission. Perhaps a knife used for preparing infected meat was used later in a scarification ceremony, although this seems unlikely since the HIV is highly susceptible to temperature changes. A more likely explanation comes from an anthropologist who recorded a custom of some of the people living near the lakes of the Great Rift Valley of Central and East Africa. In order to increase the sensations of sexual intercourse, the blood of male and female monkeys was inoculated into the pubic, back and thigh areas of men and women, presumably in an act of sympathetic magic. We know from intravenous drug addicts that such direct contact, breaking the protective skin barrier, is by far the most effective way to transmit HIV from one person to another, so this custom seems quite plausible as a possible mode of cross-species transmission. Once it is in a healthy human population transmission by sexual means becomes somewhat more difficult, but in a population subject to many other diseases, especially those producing skin lesions and ulcers, transmission is much easier. The devastating effects of HIV transmission in Africa are a later and major part of this geographic perspective on the AIDS epidemic (chapter 7).

The effects of HIV are not just changing Africa. We know from many historical records that when a disease moves through a population the consequences may be devastating. Europe lost a third of its people to successive waves of bubonic plague; over 30 million people around the world died from influenza in 1918–19; and even the annual rhythm of less severe forms of influenza, appearing in the late autumn of the northern hemisphere, and declining in the spring, takes a toll in all sorts of ways. Not only do many older people die from the annual influenza waves and their complications, but billions of working hours are lost each year. We can imagine an influenza wave moving across a country, leaving behind it all sorts of small swirls and eddies of effects on individual lives, families, and institutions. Universities are particularly susceptible as thousands of students fan out across the country three or four times a year, picking up and bringing back any "bug" around, like a great pulsating vacuum cleaner sucking diseases into the academic bag.

But the chain of effects from most diseases to other aspects of people's lives is not very long. Some daily tasks may be a bit delayed, but they can be made up later, whether you are a schoolchild catching up on homework, or a large corporation delaying the signing of a contract until the president gets over the flu. The ripples of effects are locally quite small and of short duration, and afterwards a whole society and the world of which it is a part more or less jog along as before.

But not with AIDS. Not with a slow plague with an eventual mortality rate close to one hundred percent. This is not an annual pulse giving some people a hard time for a few days, but a great, slow, global tidal wave that will take decades to peak and crash. The effects are already reaching into almost every aspect of our societies and our world, affecting the political, economic, scientific, social, religious and cultural spheres. AIDS is, quite literally, changing the world we live in, changing our planetary home, and altering the individual "worlds" that we have been thrown into. The post-AIDS world is different, very different, from the pre-AIDS world. And in each sphere the tendrils of these effects are very long. Let us see what some of them are.

3

The thin tendrils of effects

In a deep and remarkable way, thc child with AIDS is as the world's own child; the man or woman dying with AIDS is the image of our own mortality.

Jonathan Mann, while Director of the WHO AIDS Program

People who have AIDS turn the mirror we hold to their lips around to our own terrified faces.

Carol Muske, "Rewriting the Elegy," in Poets for Life

We call the HIV a lentivirus, and acknowledge a slow plague, but the impact of the epidemic has been anything but slow. Within a single decade our world has been irreversibly altered by the thin tendrils of effects that reach, directly or indirectly, into the lives of individuals and almost every sphere of collective human effort. The impact in medicine itself is the most immediate and obvious, but while the HIV moves literally through the human body it also moves metaphorically through great arenas of human life and the institutions that are a part of them – the economic, educational, cultural, artistic, social, legal, and political institutions that structure our modern world. It also shapes the ethical stances that people take within these large social structures, for ethical concern arises out of an *ethos*, a web of everyday life that informs our sense of what is right, acceptable, decent, fair and just. It is the informing thread of human life and thought that in different times and different contexts raises questions of the rights of individual people and the rights of the larger societies of which they are a part. And contained within this age-old, yet constantly new problem lie such specific questions as patient confidentiality, segregation and quarantine, the unconstrained reach of the profit motive, crime and punishment, giving and accepting, caring and denying, creating and destroying. Touching one of these thin tendrils of effects is like touching a delicate spider's web. The vibrations are transmitted through a society, and one has the sense that in human life, as in nature, everything is ultimately connected to everything else. We cannot explore more than a hundredth part of these interconnected effects, but it is important to get a sense, just a brief glimpse, of the way our world has changed.

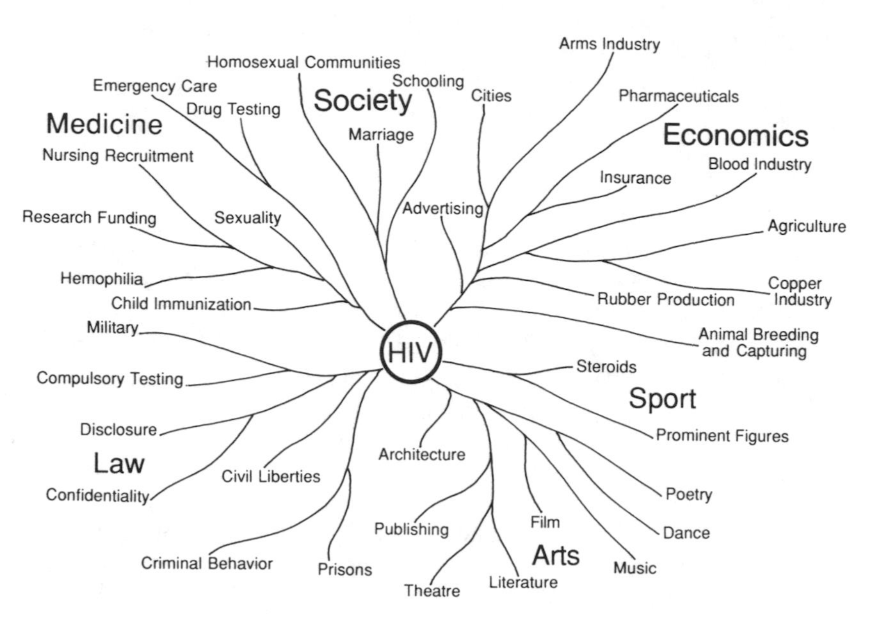

Figure 3.1 A "map" of the thin tendrils of effects reaching out and into many areas of human life. Few aspects of human life have remained untouched by the virus.

If we choose the right perspective, the HIV, that tiny blob of protein molecules, stands at the center of the web, the focal point for the map of the human condition (figure 3.1). Like any map, including the genetic map we looked at before, it somewhat simplifies and distorts. The "regions" of medical, economic, social, cultural, and other effects are not distinct, but overlap and intertwine, and they really require a multi-dimensional space to show the relations between them properly. What we have here is only a two-dimensional projection of the way the virus is influencing our lives and changing them in complex ways.

Perhaps closest to the virus itself is the region of immediate effects on medicine, medical research and healthcare. In many countries of the world the pandemic has already overwhelmed the limited resources available: parts of Africa, as we shall see, can no longer cope. The few beds available have long since been used time and time again, and "AIDS wards" are often euphemisms for concrete floors spread with mattresses soaked with the urine, blood and excrement of previous patients. Many prefer to die at home, alone, or perhaps in the gaze of yet another terrified and orphaned child. In the United States, with 360,000 people with AIDS in the first decade, a third of them dead, and 1 or 2 million infected people who are likely to convert in the decade ahead, the costs are already staggering. Compared to the GNP, healthcare costs have doubled from 6 to 12 percent in two decades, and they finally became a major political issue from 1991 onwards. It takes roughly $80–100,000 to treat a person during the average 27 months between conversion to AIDS and death, and it will cost more as drug intervention prolongs life but does not cure. In many metropolitan areas the hospital system is already overstressed, although no systematic plans have been made to expand a mix of healthcare facilities made up of conventional hospitalization, hospices, out-patient and home care. It is difficult to estimate in any really reliable way the total costs involved, but $6–8 billion probably brackets the cost of caring in 1991 alone, and the cost curve is still going up – steeply.

Which has little or nothing to do with the cost of research designed to understand and contain the HIV. In the United States alone, after an agonizingly slow and unresponsive start, AIDS research is now running at around $1.2 billion a year, spread over the medical, biological, social and behavioral sciences. Never in human history has a country spent such wealth so quickly on scientific research focused upon a single disease. At a smaller scale, research spending has also risen dramatically in Europe, where there is clear evidence of a much more focused approach, and a much better use of limited resources. But even in the "subregion" of HIV research the tendrils of effects tease out into ever finer strands and affect other areas in turn. Just as a single example, testing the effects of various HIVs and their potential vaccines requires large numbers of apes and monkeys, most of which can be used only for single experiments to isolate the effects of specific vaccines, virus mutations, or DNA alterations by genetic engineering. Animal breeding and capturing in the wild have become highly lucrative, but now raise the question of whether species, like the chimpanzee, are becoming endangered in habitats where they were once plentiful. All these efforts require huge sums of money, including the diversion of scarce financial and scientific resources that might have been devoted to research on other diseases. There are a number of other scourges that still take more human lives annually than AIDS, although they may fluctuate from year to year, or maintain fairly steady levels of death. In contrast, HIV infections are rising rapidly around the world. What is the right thing to do?

Healthcare does not simply mean allocating financial and scientific resources, but highly skilled human resources, the doctors and nurses who are at the front line of caring. It is here that the emotional stress is felt most severely. Doctors and nurses are becoming harder to recruit into the areas of medicine – mainly cancers and pneumonias – in which AIDS cases predominate. Partly it is the sense of powerlessness that many doctors and nurses feel, the sense that they can do nothing to stop the ravages of the virus and its deadly final effects. In medicine, as in many other walks of life, there must be at least a small vestige of hope to keep people going. Partly it is the young ages involved, the children and young adults in what should be the flowering and prime of their lives, and the fact that young doctors and nurses identify with them so closely. Burn-out becomes increasingly common in young nurses, who feel they can no longer go on after long periods of caring for people close to them in age and lifestyle. Despite some attempts to mechanize and "professionalize" certain aspects of nursing (for example, computers simultaneously monitoring many people for vital signs in a post-operative ward), and despite protective attempts by some to harden themselves by raising an emotional screen of reification through which people are seen as things, the emotional stresses ultimately become too much for many of them. In matters of life and death hospitals can be stressful places at the best of times. AIDS care places yet another emotional burden on those working there, and counseling services, "decompression" techniques, and group therapy sessions are becoming common ways of helping people who care to cope with their own lives within their chosen professions.

In the economic sphere the HIV tendrils are also deeply at work, and raise some of the severest ethical questions. In many parts of the world, including India and the United States, a significant portion of the blood industry is in private and very-much-for-profit hands. Anything that might reduce profits is avoided if possible, including vigorous screening of donors and testing of the blood itself. The result of such greed has been catastrophic. In India today selling blood is a profession for hundreds of thousands of poor people, usually to unlicensed blood banks, most of which have few facilities, and often less inclination, to test for HIV. In West Bengal alone, 180,000 pints of blood are needed each year, not the least because many blood transfusions mean more profits at each step up the line from donor, to blood bank, to the negotiating agent, to the hospital, and finally to the surgeon. All take their cut. Only 80,000 pints can be supplied by the state blood banks, which may undertake some, perhaps quite cursory, screening and testing. When large-scale screening is actually undertaken, professional blood donors are the third largest group infected by HIV after homosexuals and intravenous drug users.

In the early years of the pandemic the situation was not that much better in the United States, where unconscionably irresponsible decisions were taken by members of the medical profession advising the private blood industry. Testing, said the blood industry, would be too expensive (read "profit-reducing"), and anyway the chances of being infected by blood transfusions was so small that it could be considered negligible. At the time it cost the military $4.31 to test a blood sample of a potential recruit, a price that most people would not consider too high for a matter of life and death. The result at the end of the first decade was nearly 5,000 people with AIDS as a result of blood transfusions. For them the probability computations of the statistical reassurers hired by the blood industry were of little comfort.

Blood also forms the raw material for making a clotting agent used for people with hemophilia. To produce small quantities of K8, thousands of pints of blood are required, any one of which could be infected with HIV. Even if the chance of detecting the HIV with screening tests were 999 out of 1000, the probability of an infected pint getting through 2000 donations to contaminate clotting injections would be $1 - (.999)^{2000}$, or virtual certainty. The result is that nearly 2000 hemophiliacs in the United States alone have already been infected, and the tragedy does not stop at the borders. This is a highly profitable industry, and exports were brisk in the early eighties. Hemophiliacs from Japan to Denmark have been infected by the contaminated shots and blood transfusions.

Where the profit motive is strong, ethical considerations tend to be pushed aside or dismissed with comparative ease. One glossy magazine of the international arms industry, whose advertisements touted the latest in everything from armored personnel carriers to tanks, fighter aircraft, and other efficient means of slaughter, noted in an article that Africa was clearly a continent upon which the astute arms dealer might keep an entrepreneurial eye. As we shall see, the armed forces of many East and Central African countries have extremely high rates of HIV infection; for example, half of the Ugandan Army, and most of the officers in the Ugandan Air Force. When the political and military leadership of a country collapses in a slow plague, all sorts of opportunities arise in the international arms trade as new strongmen come to power. New toys for the boys means that an honorable arms dealer is not without profit in his own country.

Two industries most directly affected are the pharmaceutical and insurance companies, one facing potentially huge profits, the other equally large losses. Those who developed AZT, the drug most used both before and after conversion to AIDS, recovered their research costs in the mid-eighties, and ever since have made profits of hundreds of millions of dollars each year. At $8,000 for a year's treatment, and 100,000 people being treated each year, it is not difficult to see that stockholders can anticipate a reasonable profit for their bold risk-taking. Pentamidine sprays used to retard the development of pneumonias may be equally profitable. At $99.45 a dose, which works out at $1200 for a year's supply, the American company can expect to be well in the black, even though a French company sells the same thing for $28 and does not appear to be going bankrupt, perhaps because sales to Mexico are brisk, because many are resold at $58 over the border to the north. The entrepreneurial spirit is obviously alive and well, and many accept that this is the way the system is meant to work.

For the insurance industry the shoe is on the other foot; you increase profits either by increasing sales or reducing costs. When a slow plague comes along, with millions of people seeking medical, job loss or life insurance, the best way to reduce costs is not to insure those who might shortly make large claims. This means testing for HIV before insurance is issued, and declining to insure those who are seropositive. In Eire, to take but one of many examples, proof of seronegativity is required before a policy will be issued, and in Britain the industry has recommended testing, or including cancellation clauses, for HIV infected people. In a nation like the United States, the private insurance companies have long opposed any proposals for a scheme for national health insurance, saying that they were by far the most capable people around to handle this lucrative source of revenue. But like many institutions extolling the merits of the market and free enterprise, they are quick to offload on to the government and the taxpayer any potential burden that could increase their costs, thereby reducing the profits skimmed from the more than $200 billion in premiums and receipts taken in each year from annuities alone. Looking pathetically noble while lying through your teeth is one of many tactics designed to raise sympathy. By the end of the decade, AIDS care was only about 2 percent of the nation's health bill, but major lobby groups were still citing a cost of $147,000 per claim in public, while their own in-house studies showed an average claim of $36,159.

But the effects of the HIV on the insurance companies generate other tendrils in their turn. Many moneylenders require a borrower to take out a personal life insurance policy to secure the loan, and many people fear that if they are found to be infected their records will not be confidential. As the word about a denied policy gets around, or even generates gossip and speculation, employment and other forms of discrimination may appear. The result is a total contradiction. While the medical profession is pushing voluntary testing in order to counsel more socially responsible behavior, the "voluntary", take-it-or-leave-it testing required by the insurance companies drives potential people at risk away. In many homosexual communities the fear of losing insurance coverage was the reason most people cited for not getting tested at all.

But testing itself is only one of a number of lucrative industries that have sprung up or grown around the HIV. In the 1980s, Illinois, Louisiana and Texas required HIV testing before issuing a marriage license, using the precedent of the required Wasserman test for syphilis. Some firms in Illinois charged over $200 per test (today the cost is about $1), and all they succeeded in doing was to drive young couples over the border to adjacent states to be married. State laws have a distinct geographical reach. The whole program collapsed and was eventually canceled on the grounds of being totally ineffective. Other industries have also seen a boom, a point not unnoticed by the stock markets, particularly the most volatile ones like Tokyo. If you want to, you can put together a whole investment portfolio of AIDS stocks, mainly companies engaged in biotechnology and genetically engineered drugs, and in this way make a profit from other people's misfortune. With the AIDS crises, investors pushed the stocks of Japanese condom companies, already high because the contraceptive pill is illegal in Japan, up 192 percent, while in the United States sales rose 94 percent in four years, but then leveled off. Worldwide there has been a greater demand for natural latex, not just for protective gloves, but for condoms and other accoutrements of modern sexuality, such as rubber dams for lesbians who wish to engage in cunnilingus. This has already produced a modest revival of the rubber industry in countries like Guyana.

A chain of effects from the HIV to changes in protective behavior, to increased condom sales, to rising demands for latex, to increases in the rubber industry in Guyana, reminds us that what happens in one place is seldom independent of what happens in others. At the same time, the HIV pandemic circulating around the globe "touches down" with particular destructive force at some places, rather like tornadoes appearing out of a more general pattern of global circulation. Over one-fifth of all AIDS cases diagnosed in the United States have appeared in New York City, and a number of thoughtful people now feel that the increased stress jeopardizes the city's standing as a business and financial center. There is an acute shortage of hospital beds, yet good healthcare is becoming an increasingly important factor in companies' locational decisions. Businesses might be bribed by the metropolitan government with tax breaks and tax payers money to hang around a little longer as a source of employment. But costs of AIDS care to the city will soon be close to $2 billion, with nothing but an already well-eroded tax base to draw from. This means higher taxes, higher health insurance (or healthcare benefits eroded by discrimination), and still further cuts in other services in a metropolitan area in which some boroughs have already had the rug of fire, police and basic social services pulled out from under them. One former head of planning in New York, not the world's most enviable job, speculated "Are we going to be stepping over bodies, like Calcutta?"

In the world of the arts the impact of the HIV has been particularly poignant, not simply because the thin tendril has had an informing and catalytic effect upon artistic creation, but because the homosexual community has been particularly well-represented in the world of artistic creativity and culture. Taking the first shock wave, the first "touch downs" of the AIDS tornadoes, many highly creative people throughout the arts have been lost. It is a community that has shown exemplary care and compassion for infected people, and the proceeds from many artistic productions and works of art have been given to help those afflicted. Some contributors had already converted to AIDS themselves, but with great courage they continued to create for others. Philly Lutaya, a popular musician and composer, performed for the sake of more effective education in his native Uganda until a few days before his own death, and from Nigeria to the Bronx the lines sung or rapped in popular music appear to have a greater educational impact upon teenagers than all the millions of dollars spent on what is ostensibly, and too often euphemistically, called "educational research." Some tape cassettes of popular music even have educational messages about AIDS, although the lyrics of some of the songs extolling the joys of uninhibited sex might appear to contradict the message on the cover.

Literature, poetry and plays have all mirrored the terrible human consequences of the pandemic, including searing accounts of the drawing near of death, and the attempts to express and share a sense of loss of those loved. Plays like *The Normal Heart, As Is, Safe Sex* and *Coming of Age in Soho* have played to large audiences, whose experiences with such theater often leave them more thoughtful, understanding and compassionate. Perhaps more than any other, the world of dance reflects the tragic losses, for the pandemic has taken some of the brightest and best in classical and modern dance, choreography and spectacular dance production – like Michael Bennett of *Chorus Line*, and the principal dancer Charles Ward of American Ballet Theater. Modern dance itself reflects the theme, as in *Tainted Love*, a production of the London Contemporary Dance Theatre that also toured the United States. As one tragic loss after another is reported, one is reminded of the huge holes in the fabric of the arts caused by the slaughter of young men in the flowering of their youth during the First World War. How much have our own lives been diminished as a consequence by the "if only" and the "might have been." Where have all the flowers gone . . . long time passing?

Even the world of film has seen some small shifts in its usual emphases on steamy sexual encounters with multiple and highly nubile partners. James Bond in *Dragnet* has suddenly become remarkably monogamous, especially when he discovers that his box of condoms is empty so that his potential blonde partner in pleasure must continue to yearn for a closer experience of his manly, but presumably fleeting, affections. Though hardly qualifying as artistic productions, pornographic films are now appearing in which safe, i.e. condom-protected sex, is the theme, a message also taken up in the genre of homosexual "literaturc."

No area of aesthetic creativity appears untouched or unmoved. Architecture has responded in a remarkable way, designing housing for people with AIDS within a larger project called Vacant Lots, sponsored by the Architectural League of New York. On city-owned, or cheaply purchased lots in often run-down areas, a home can be built in two separate units with a garden between. Using prefabrication where possible to cut costs, individual rooms, each with its own bathroom, adjoin dining, kitchen and recreational rooms which serve all members of the community. There are also rooms for nursing and treatment, but the projects have been designed to be as self-sufficient as possible for people becoming increasingly handicapped, and they are small enough to minimize their impact on the local area and so be accepted by the neighborhood. Not only is an older, often much-dilapidated housing unit renewed, with all the possible demonstration effects spilling over into the local neighborhood, but the public costs can be weighed against long-term stays in increasingly overcrowded and stressed hospitals. Whether writing a sonnet, or creating a home for people, one always works within constraints whose forms may themselves induce and shape a creative response.

Where do the tendrils of effects end? The world of law has been profoundly affected as vexed questions of individual rights versus the rights of society have emerged in countries whose democratic ideals are rooted in and guarded by extensive bodies of legal opinion. All claims to the contrary, a body of law is neither given by divine intervention nor by nature. It is always a human construct, a part of a society emerging from its prevailing ethos, and it can only respond to the slow and conservative social dynamic in which it is embedded. Despite blind Justice, with her sword and scales in hand as a constant reminder, we tend to forget that the courts enter those areas of human life where a sense of injustice is felt, where people feel something is unfair, where "it just ain't right." When HIV moves through a population the sense of injustice may be high. In the first decade, in the United States alone (admittedly a chronically litigious society), nearly 1000 cases had been filed in federal, state and local courts, involving the right to educate children about the dangers of the virus; the safety of the blood supply; discrimination in jobs, housing, insurance and education; the responsibility of a government to monitor a deadly disease by surveillance and testing; the right of an individual to privacy; the criminality inherent in passing on the virus to another person; and the liability of a commercial company doing its best to produce vaccines that initially are always experimental.

When does the right of a rape victim become stronger than the rapist's right to confidentiality? When can a company fire someone who tests seropositive? When are you in your "right mind" to make a will if AIDS-related neurological disorder is already apparent? Does a wife have the right to know if her husband is infected? Can we lock up and effectively quarantine those who declare "I'm gonna take as many with me as I can!"? Are death certificates public documents, or may they be withheld from the "prurient interest" and "idle curiosity" (two phrases actually invoked by state court opinions) of the mass media? What are the rights of people in prison – for healthcare, treatments, protective isolation, counseling, and so on? Is a doctor responsible for not reporting an HIV-infected patient? Is it criminal, in both the strict legal and everyday use of the word, to withhold information from teenagers who might otherwise die? At what level of aggregation may AIDS cases be reported to researchers trying to understand and predict the future course of the epidemic? We shall look much more closely at this question later (chapter 13), because it is essentially geographic.

As the HIV works its way through society, like a floodtide working its way along the channels, creeks and tributary filaments of a salt marsh, so it touches each and every one of us. There are few people left in Africa who do not know someone, or two, or three . . . who have died of AIDS, and such immediate and tragic knowledge is becoming increasingly common in North America and Europe. As the pandemic comes closer to us personally, so it changes, sometimes quite radically, the ideas and views we have of ourselves and others, and the conversations we are prepared to take part in. For many people around the world, sexuality in all its forms has been regarded as an intensely private matter. It was not customary to discuss making heterosexual love at the dinner table, let alone the possibilities of homosexual anal sex. The ethos, the worlds into which most people had been thrown, made such topics not simply unsuitable but unthinkable. For many they still are – under any circumstances – and the silences still prevail.

But they are deadly silences, silences of death. We have no cure, no vaccines. Education is the only thing available to halt the pandemic, and slowly, all too slowly, we are realizing that education means talking, answering questions, bringing things into the light of conversations and dialogues that would have been unthinkable a few years ago. Parents *must* talk to their children; teachers must talk to pupils; lovers must talk to one another; students must talk to students; and politicians must talk to constituents. And, perhaps even more important, constituents must talk to their representative politicians. Kentucky and Norway have about the same number of people. Kentucky has the lowest rate of high school graduation, and the highest rate of teenage pregnancy, in the United States. By the first decade it had over 700 recorded AIDS cases (perhaps 1200 under revised definitions), 30 percent of them under 30 years old, which means that they were infected in the teenage or early adult years. This is just the beginning, and there is no end in sight. In many rural high schools in Kentucky the AIDS epidemic cannot be discussed, let alone such nasty things as sex and condoms that might embarrass a Southern Belle. Meanwhile the Good Old Boys in the state legislature often let the AIDS funding run out for a while just to teach "them" a lesson. In marked contrast, in Norway, as in other Scandinavian countries, superb, open and responsible sexual education, undertaken by trained professional teachers, has been the norm for over half a century. There is some strong evidence that there the HIV epidemic has been turned around, and may, like the gonorrhoea rate, be reduced to almost nothing.

Talking works, and it can take place. Not long ago I was asked to talk to pupils of a nearby girls school from the seventh grade up who had just completed a month's "research project" on the pandemic. "What," said a proper but very forthright young lady from the eighth grade, "What is the chance of getting the HIV from oral sex?" In a lecture hall, with perhaps 200 young women, there was not a smile or snicker of embarrassment. This was serious stuff. "At the moment," I replied without blinking an eyelid, "we only have one well-documented case, published a few months ago in *The New England Journal of Medicine*. But it is obviously possible, so why risk it if you're not sure?" Perfectly satisfied with the answer, we went on to discuss other forms of transmission, including anal sex. There was no prurient curiosity, no sense of daring in raising such questions, and in latino cultures, where virginity is still strongly pressed for upon marriage, it is not a form of intercourse that we can pretend does not exist as a possibility for young people in the intense dawn of sexual awakening and experimentation. What had been created in that girls school was an atmosphere of openness and seriousness in which these questions could be asked, discussed and answered. And almost certainly in the years ahead lives would be saved.

So our "map" of the effects of HIV on all spheres of human life is not as fanciful as it first seems. The tendrils may not be as physically immediate as the virus's thread of glycoprotein gp41 that connects and invades a human cell, but they are there nevertheless and are often just as devastating. At the same time our graphic picture is not complete. To think like a geographer, to recall that viruses and human beings exist in space as well as time, we have to realize that these effects are dynamic in time and are played out in specific, yet always interconnected places (figure 3.2).

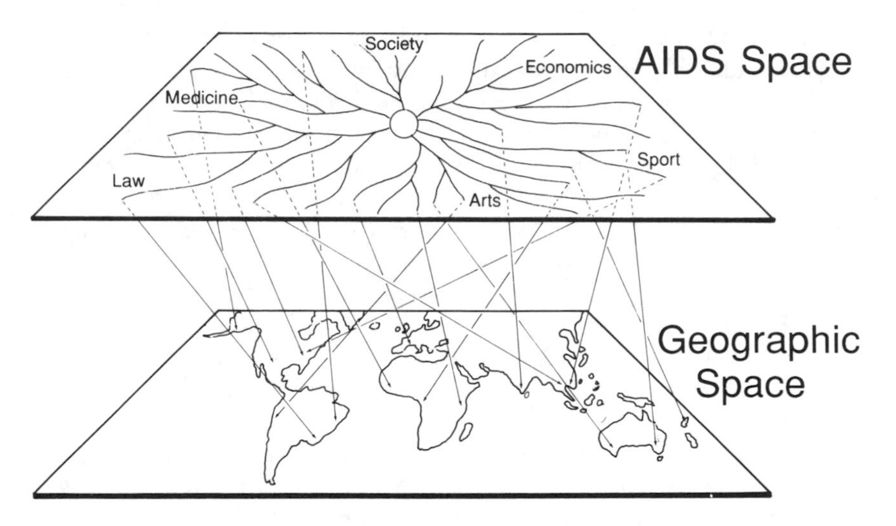

Figure 3.2 Projecting the map of HIV effects onto the more familiar map of the world. This one-to-many mapping reminds us that the thin tendrils of effects work their way through human communities in real spaces and actual places.

What we really need to do, in something that a physicist would call a thought-experiment, is project our map of effects onto the concrete configurations of our planetary home in all its marvelous human variety and complexity. We can only do it schematically here as a sort of graphic *aide-mémoire* to help us remember that a pandemic always has a complex geography as well as a history. Significantly, in projecting our rather strange map of "AIDS space" onto the more familiar geographic map of the world, we are using what a mathematician would call a one-to-many *mapping*, taking events from one space or domain into another, relating the things and events in an abstract AIDS-space to the space of real locations where people live and die and have their being.

But no matter what spaces we think in – abstract or concrete, conceptual or geographic – they are never unformed, amorphous and unstructured spaces, like Isaac Newton's cosmic wastepaper basket into which the deities casually tossed the universe. In human geographic spaces, things, events, places and people are connected together, sometimes directly, sometimes by long strings of intermediate connections. Quite tragically, it is those connections between things and people that form the structures upon which the HIV exists and is transmitted. To understand what is really happening we need to think just a bit more about this business of connections forming structures, and how the structures must be in place for other things to "have a home" and "find their way about." Both good geographic ideas.

4

Sex on a set: a backcloth for disaster

Safer sex is the only option for women sex workers like everyone else . . . the fact that money changes hands is neither here nor there. The virus doesn't travel on dollar bills.

Cheryl Overs, Victoria [Australia] Prostitutes Collective

Sometimes I think God was not entirely serious when he gave men the sexual instinct.

Father Jean, in Graham Greene's The Burnt-Out Case

The AIDS pandemic is so humanly devastating that it may seem initially almost obscene to stand apart and take an abstract and apparently unfeeling view of it. If, in the pages ahead, you feel that the discussion of geometries and structures gets too abstract and rarified, if you catch yourself saying "Who is this guy? Just another academic in his ivory tower who doesn't realize this is happening to real people?", I shall not only understand but sympathize to a degree you might not suspect. If we want to think a bit more precisely about what is going on, we have to acknowledge that any perspective on the human condition that requires us to stand apart, to detach ourselves from an immediate concern for an individual or group of people, constitutes a reification, what one geographer has termed the "thingification" (*res* = thing), of the human being. Human beings are not things, and we slip into reifying habits of thought only at our own deadly peril. When people are truly reified, when they are thought to be just things, just abstract elements of sets, we are in the realm of the Nazi deathcamp, the Stalinist gulag, the My Lai massacre, Idi Amin's prisons, and the world of the unremorseful psychopathic killer.

That phrase about treating people as "just abstract elements of sets" should warn us that reification frequently means mathematization, and it is true that the fairly abstract ideas we need can be treated in very sophisticated algebraic ways within combinatorial topology and other quite rarified branches of mathematics like hypergraph theory. Nevertheless, the basic ideas are simple and intuitive. All we ask is: what does a deadly virus need to exist, survive and be transmitted? The answer is equally basic: it needs a "home," a place where it can live and reproduce, and various sorts of connections over which it can be transmitted and so spread or diffuse. Many diseases do not need people as "homes": influenza can exist in pig populations; Lyme disease is found in ticks, deer and whitefooted mice; cholera needs water, and is quite happy in shellfish eaten by people; and the tetanus virus is found in soil. Even the HIV crossed over from monkey and ape hosts. On the other hand, certain diseases seem to be quite human specific, which is why we were able to knock smallpox out of existence with vaccines, and why we are going after measles now.

The HIV, particularly in its mutated forms, is virtually human specific. The chances of infection from ape and monkey populations today are vanishingly small, because few people live in close proximity to wild mangabey monkeys or chimps. The HIV needs people to exist, and it needs connections between people to move from one person to another. If we had simply an unconnected and unstructured set of people (figure 4.1), a person infected with HIV would eventually convert to AIDS and die, and the virus would also disappear. But people are seldom unstructured sets, they are connected by relations – all sorts of relations – in both the common and mathematically rigorous senses of the word. *Relations* connect elements of sets and thereby form structures, and it is these connected up and structured sets of people that we are going to call *backcloths*. It is on a human backcloth that a virus exists as a *traffic*, and it needs the backcloth of connective tissue to move from person to person as *traffic transmission*. As we shall see, to stop the HIV traffic transmission you have to break the connections and so fragment the backcloth.

What sorts of relations might structure a set of people into a backcloth for HIV? The major one is the sexual relation in all its variety of forms. If it takes place within monogamous marriage, either hetero- or homosexual (figure 4.2), then the traffic of HIV (perhaps originally from a blood transfusion) can be transmitted only to the partner. The backcloth remains highly fragmented, and there are still many gaps or obstacles the HIV cannot jump across. Both partners may eventually convert to AIDS and die, but again the virus disappears with them. Notice that polygamous or polyandrous marriages structure people in much more connected ways (figure 4.3), and a single infected man or woman can infect all the sexually connected partners. We would expect

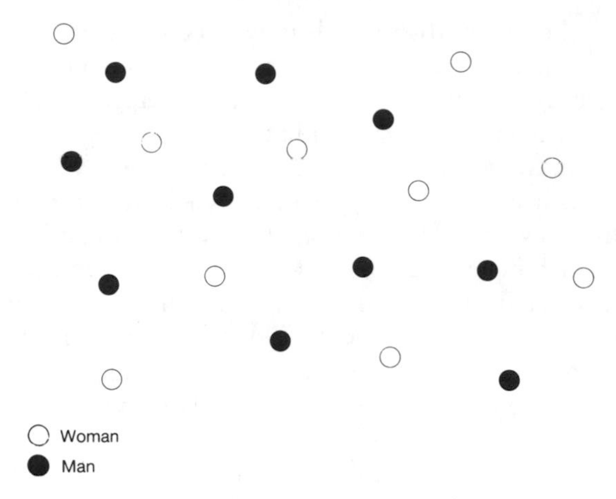

Figure 4.1 *A set of people (males solid circles, females open circles), unstructured in any way by relations between them. The HIV might live on one or two people in the set, perhaps transmitted by a blood transfusion, but could not be transmitted to others.*

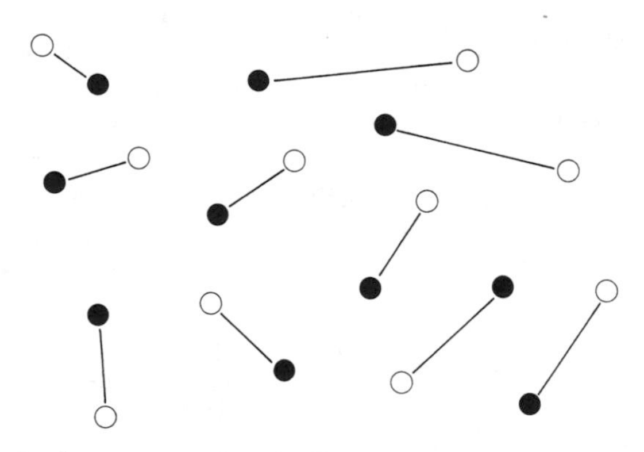

Figure 4.2 *A set of people structured by the monogamous marriage relation. If an HIV exists on one or two people in this structural backcloth it can only be transmitted to the marriage partner, and then disappear upon the death of both of them.*

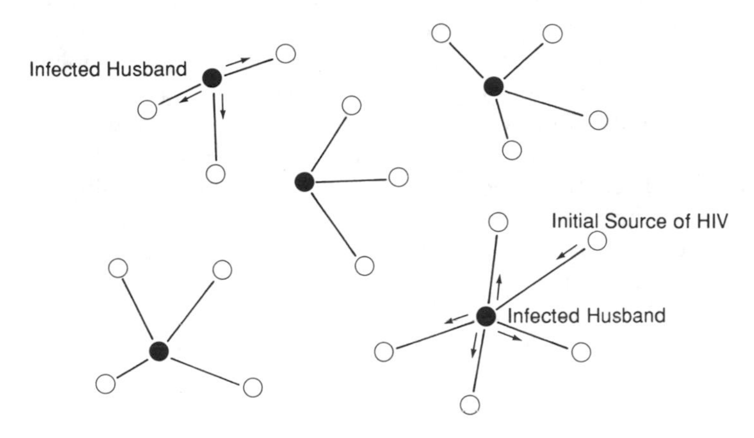

Figure 4.3 A set of people structured by the marriage rules and relation of a polygamous society. The sexual connectivity of this backcloth allows for a much wider transmission of HIV.

the HIV to move more quickly in societies with multiple marriage partners, including societies with high rates of divorce and remarriage, or so-called "serial monogamy." And if a society had the custom of a man caring for the wife or wives of a dead brother, and of these wives conceiving children to indicate to all that he had done his duty and filled his fraternal responsibility, then even more connections would structure the backcloth (figure 4.4). This is not uncommon in African cultures, but if the brother had died of AIDS there would be a strong

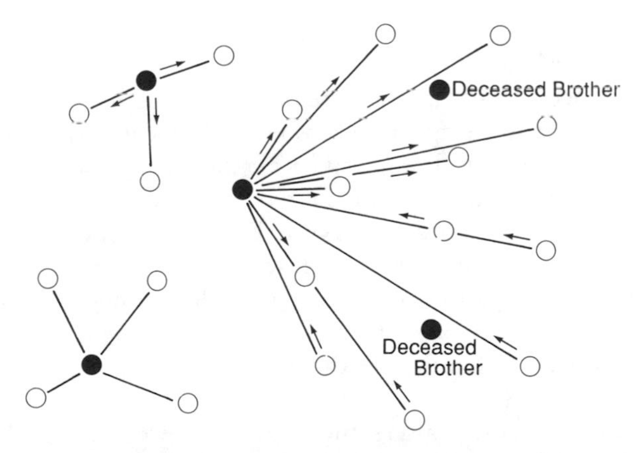

Figure 4.4 A set of people structured by the polygamous relation and the obligation to look after any wives and children of brothers who are dead. The HIV can now be transmitted to all members of the enlarged family.

possibility of all the wives being infected too, because the HIV would now be transmitted over the culturally restructured backcloth.

Sexual relations are not confined to monogamous or polygamous marriages. Different cultures allow or forbid, encourage or discourage, sexual relations, and so produce different backcloths with different possibilities for transmission. These backcloths themselves are the product of highly complex social rules, including religious prohibitions, and like many other aspects of human life they may well change under the impact of new possibilities opened up by technological advances. The contraceptive pill radically altered sexual relations, producing many more, often multiple connections that formed long chains over which the HIV could move (figure 4.5).

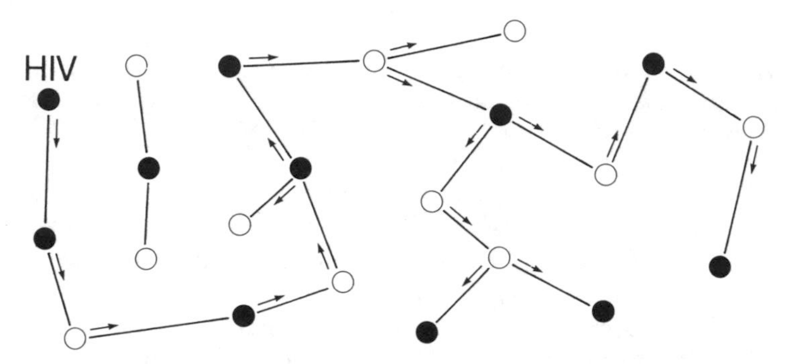

Figure 4.5 A set of people structured by multiple and unprotected sexual relations, allowing the HIV to be transmitted over a tightly connected backcloth.

In many parts of the United States, 70 percent of the teenage population have experienced sexual intercourse before leaving high school, and half of them use no contraception whatsoever. We know the HIV is firmly entrenched in this population, since an increasing number are testing seropositive, and many now in their twenties were clearly infected as teenagers. In major cities like New York, Washington and Miami, infection in teenage runaways is close to 10 percent, mainly through male and female prostitution. It is even higher in many Brazilian cities where young runaway boys, the "Princes of the Night," dress in styles that signal their availability on the street to potential customers shopping around in cars.

Prostitution radically alters the connectivity of a backcloth (figure 4.6), as multiple partners connect to just a few nodes of intense potential transmission. It is for this reason that prostitutes (both male and female) should be the first to be reached with education at the start of a

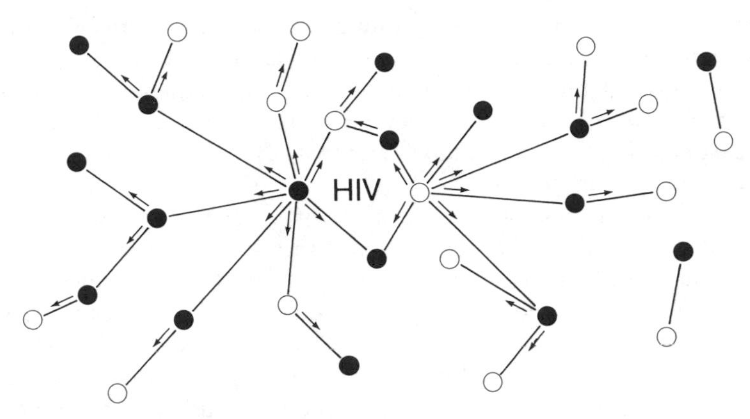

Figure 4.6 A set of people structured by multiple sexual relations, including a male and female prostitute, both infected by the HIV. The structure of this backcloth allows for rapid transmission of the HIV unless structure-breaking condoms are used.

potential HIV epidemic, and that everything should be done to help them practice safe sex. We shall see later some of the devastating consequences when this basic fact is disregarded. For a group of heterosexual men and women of any size, the average number of sexual partners is always the same, so the opportunities for infection are also equal, something that most people do not find intuitively obvious

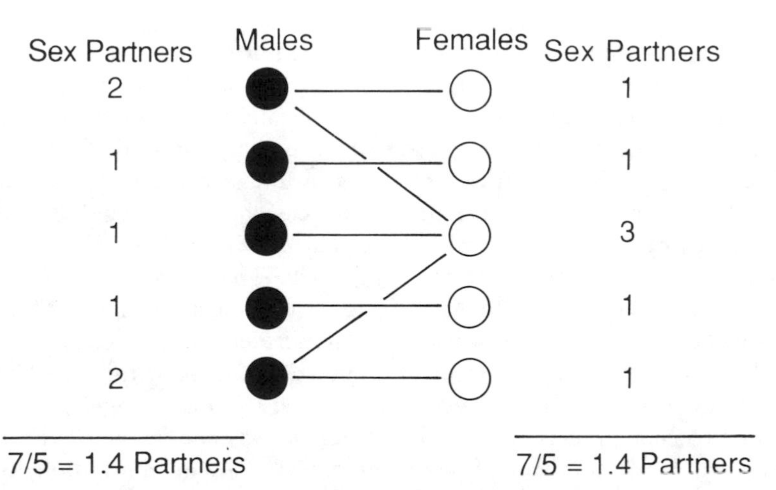

Figure 4.7 Five couples structured by their sexual relations, showing that the average number of sexual partners is always equal.

at first. Suppose we have five couples (figure 4.7), with two purely monogamous partnerships, while the female of one couple also has sex with two of the other men. If we list the number of sex partners by each man and woman, and take the averages, they will always be the same, something that immediately warns us that averages may well hide more than they disclose.

If we think in these structural terms, it is easy to see why in many parts of the world the homosexual and intravenous drug communities (sometimes overlapping) were the ones to take the first shockwave of the pandemic. A hypodermic needle shared by drug users literally constitutes a structuring relation on an otherwise disconnected set of people, which is precisely why clean needle programs are so vital (truly a matter of life and death) to prevent HIV transmission. In the Netherlands, a country that treats drug addiction as a medical problem, and where clean needle programs have long been supported, the infection rate is 8 percent, which is still appallingly high in absolute terms. In Italy, where addicts are rounded up by the police and imprisoned, and no clean needle programs are in effect, the same rate is 70 percent. Needles can be highly structuring relations on sets, and they are not confined to addict communities. One of the worst outbreaks of HIV in the Soviet Union was in a children's hospital in Elista, where the HIV was transmitted from one infected child to 61 others by a contaminated needle. Today there are grave worries about the child immunization program sponsored by the WHO in countries that have neither the means nor the medical discipline to sterilize needles after every use. An inoculation designed to protect a little child may carry HIV with it. Shared needles also structure sets of athletes taking steroids to increase their muscular bulk and strength, but the HIV is no respecter of fitness or physique.

In many homosexual communities, especially those in large cities, the sexual relation may structure a set of men so tightly that the HIV can reach every "element of the set" in just a few steps. Some of the finest epidemiological work on the pandemic was carried out in 1981, when some sort of agent of transmission was already suspected, although the HIV was not actually identified until 1982. The sexual contacts of 40 men who had converted to AIDS were carefully traced (figure 4.8), and led to the probable identification of Patient Zero, later identified as an airline steward with international connections. Even within the United States the trail of structuring connections led from Los Angeles, to New York, to San Francisco, and five other states, and most of the men reported large numbers of sexual partners. Patient Zero thought he had had an average of about 250 different male partners each year during the previous three years, although he could remember the

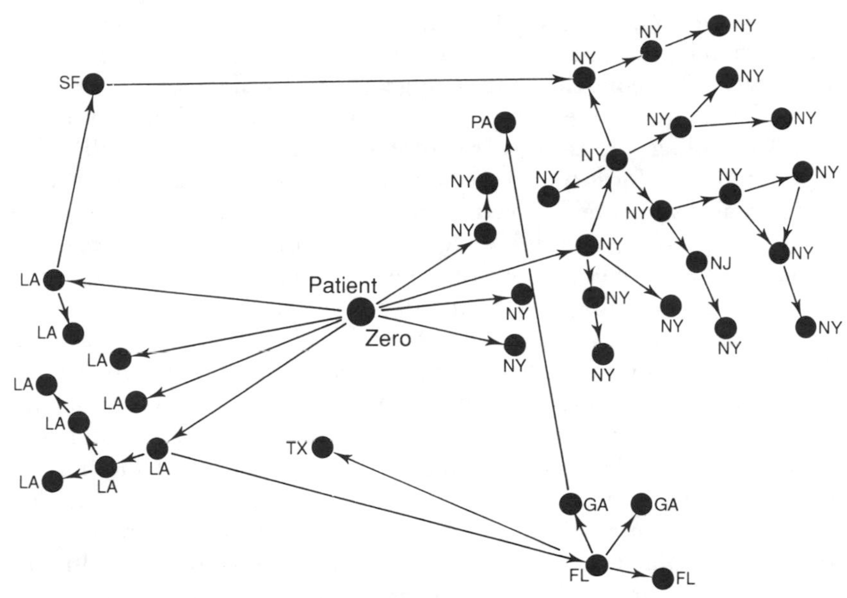

Figure 4.8 The original tracing of 40 homosexual men and their sexual contacts, leading to the probable identification of Patient Zero. It was these tracings through the Los Angeles (LA), San Francisco (SF), New York (NY), Florida (FL), Georgia (GA), Texas (TX), and Pennsylvania (PA) communities that led to the suspicion that AIDS was transmitted by an infectious agent – the HIV, identified a year later.

names of only 72 of the roughly 750 liaisons. Nevertheless, eight out of the sample of 40 were infected by him, and from this node the tightly structured lines of connection quickly transmitted the HIV to others.

Whether Patient Zero was the initiator of the pandemic in the United States will never be known with certainty, and in any event it is of little importance since at a global scale airline travel virtually guarantees rapid transmission, and many others would have unwittingly carried the virus across the Atlantic River. It is reasonable to assume that he was infected in the early seventies (his own symptoms first occurred in December, 1979) from homosexual relations overseas, and then spread the virus to perhaps scores or even hundreds of partners. Notice, in this quite concrete example, that we have moved through a series of geographic scales: from the "local" geographies of certain districts in Los Angeles, San Francisco and New York experienced by individuals; to the national scale of the United States, where we can think of the HIV spreading from state to state; to the global scale of airline travel, which is the major connector of sets of countries into an international backcloth over which the HIV is transmitted. Not surprisingly there is a direct relationship between the number of international airline connections a country has and the spread of HIV. We shall have to think much more carefully about this matter of geographic scale later.

We have already seen how blood transfusions, and blood products like K8 clotting injections, form routes of transmission for HIV, but blood may also form the connection for the tragic cases of mother-to-child transmissions and infection in health workers, whether they are surgeons operating, dentists doing intrusive work, or paramedics under accident or other emergency situations. An infected mother may transmit HIV either during birth or later while suckling the child, and conversion to AIDS is particularly rapid in small children – an average of only six months. The chance of infection for health workers is small but finite, and it is worth noting the wide disparity between the way AIDS patients were first approached in the United States and in a country like Sweden. For the most part, no protective clothing or gloves were used in the United States, since it was assumed that even the contact required during a medical examination was unlikely to result in transmission. Although over five million people in health and related fields routinely come into contact with blood, it was only after the first decade of the epidemic that legally-enforceable regulations concerning the provision of gloves, masks and smocks were put into effect. In marked contrast, the Swedish medical profession immediately treated HIV as an infectious virus, and required all health workers to use protective clothing and gloves. I suspect that both professions would claim that they had adopted the right approach, the former claiming the Swedes over-reacted, the latter deploring an apparent casualness in the face of a new virus whose transmission properties were largely unknown. Sweden today still has no health workers infected from treating people with AIDS; the United States has scores. But then the United States has more than 300 times the number of people with AIDS. Who was right?

The matter of protective gloves and clothing raises the critically important question of the possible ways the HIV can be halted. In terms of our structural backcloth the answer is naively obvious, even though it may be politically, economically and culturally impossible to put into effect. It is quite simply to break the structures of transmissions, to change the relations on the sets, and fragment the backcloths so that the HIV cannot jump across the gaps. Obviously the most desirable way is to destroy the elements of the set forming the potential material for a backcloth. Then the HIV as traffic has nowhere to live and disappears too. This is precisely what a vaccine would imply: it would remove a population at risk, in just the same way that the smallpox vaccine finally snuffed out the last stronghold of this former scourge. A bit closer to home, we know that people who have had a certain strain of flu are no longer at risk once they get over it.

The traditional way of stopping a disease, even though it is such an anathema to many people today that it becomes unthinkable, is to isolate infected people physically so that they cannot infect others. This is quarantine, for which there is ample precedent in human history, from ships with yellow fever flying a yellow flag (still the sign for quarantine in the international code of the sea), to villages trying to isolate themselves during the great plagues. There is also enforced medical quarantine for smallpox, cholera and other infectious diseases, a degree of isolation supported by the laws of many countries. It is the method enforced by a few nations for HIV, notably Cuba, Bangladesh, and the state of Goa in India. It is also strongly "suggested" in Sweden for most people needing hospitalization who are referred to a central and specialized facility on an island near Stockholm. Anathema or not, it is a nice exercise in social ethics to ask someone what they would have done if they had been Minister of Health in Cuba during the period that thousands of young men returned from Africa. Cuba is the only country to attempt the testing of all its citizens for HIV, and it isolates infected people in a special hospital at Los Cocos south-west of Havana for six months of medical care and education. After that time, people are allowed to visit family and friends, but with a clear understanding of what unsafe sex would mean in terms of a possible death sentence for others.

These are truly exceptions that only prove an almost universal rule. Most countries do not quarantine, and it is worth asking why. The reason usually given is that quarantining or compulsory testing would have exactly the opposite effect to what was intended. It would discourage potentially infected people from coming forward if they knew they would be quarantined, or even if their infected state were disclosed beyond a small number of medically trained people. There is also a sense in most democratic societies that "it just wouldn't work." The virus is already so widespread, even though we try to avoid thinking about where it is, that national testing alone would be infeasible, and quarantine impossible. How are you going to quarantine between one and two million people already infected in the United States?

But there is another reason, and it has to do with the way we perceive the immediacy of infection and its danger to us. To think about this we need a bit of graphic help (figure 4.9). Like many processes in nature, an epidemic is self-limiting. In other words, it does not go on forever, but gradually burns itself out. "Burning itself out" is not just a vivid euphemism, but actually points to what happens: the "fuel" for an epidemic is always a population at risk, and this is consumed by the viral "forest fire" until either everyone is dead, or they have recovered with antibodies in their systems which give them future immunity. When we plot the course of an epidemic over time, recording the history of the disease (but ignoring the geography for a moment), we typically get an S-shaped or logistic curve.

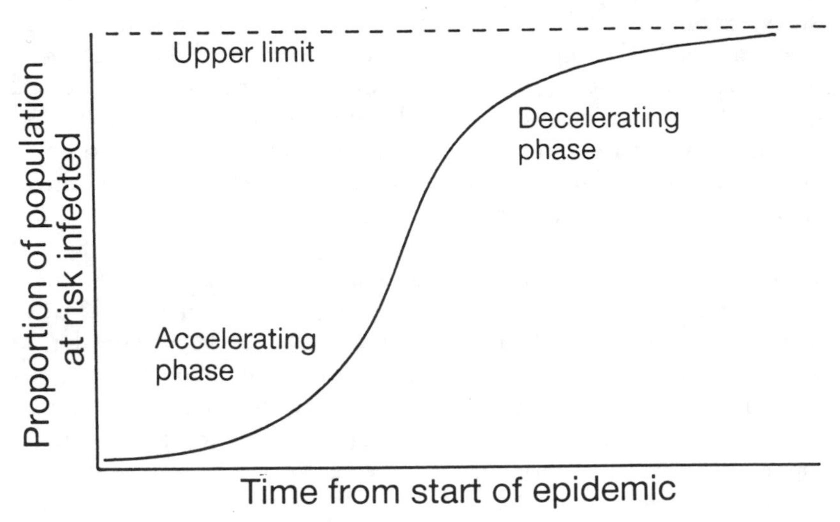

Figure 4.9 A typical S-shaped or logistic curve describing the course of an epidemic over time. An initial period of acceleration is followed by a period of slowing down as the population at risk dies off, or recovers with immunity to future attacks.

While specific examples may be more complex – a virus like the HIV may go through rather different populations at risk and produce a series of linked S-shaped curves – we usually see an initial stage of slow growth, followed by a period of acceleration as the disease catches hold. Then, gradually, as people die or become immune, the epidemic starts to slow down and gradually reaches an upper limit and stops. Vaccines, of course, can squash that upper limit of people at risk down and down until no person can be infected.

The S-curve is the most general description of the course of an epidemic, and it can be derived from even simpler and more fundamental processes, and dressed up with all sorts of mathematical paraphernalia. But for our purposes we only have to note one thing: while the general shape of the logistic function often describes quite well what is going on over time, the specific form can be quite different depending on the nature of the disease. For example (figure 4.10), a highly contagious disease (Lasa disease, measles, typhoid, etc.), with a

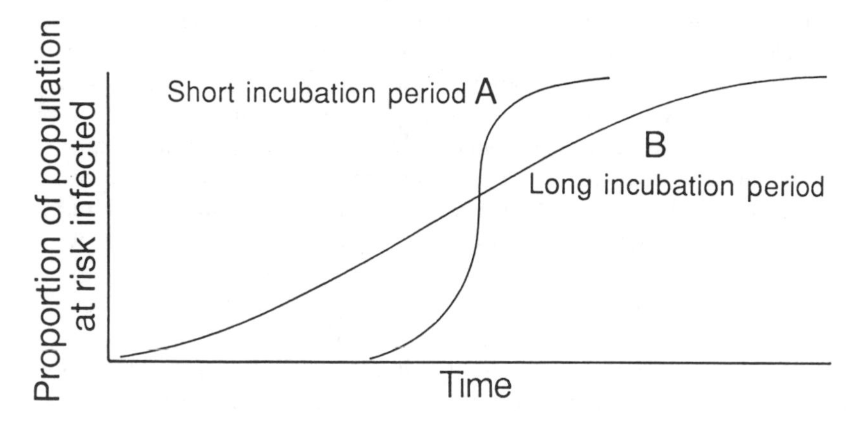

Figure 4.10 Two logistic curves describing the course of two epidemics over time. Curve A describes a disease with a very short incubation period; curve B is a disease with a long incubation period, perhaps HIV/AIDS.

very short incubation period between infection and the first signs of symptoms, will take off very quickly, and "burn off" the tinder-like fuel of the population at risk in a short time (curve A). We could almost call it a "flash-epidemic," and in the face of these many people have quite sensibly opted for quarantine. You isolate the infected people as quickly as possible, disconnect the connections, and try to limit the epidemic to as few as possible. There is often nothing else you can do.

But what about a slow plague, a plague not with hours or days but years between the first infection and later symptoms – like HIV? Then the logistic curve will describe an agonizingly long history, not over months but decades (curve B) – unless, of course, a vaccine can be found in the meantime. With something moving so slowly but surely we react in a quite different way, not the least because it takes a long time for most people ever to see or know someone with AIDS. This becomes a critical problem when we try to educate young people – teenagers and young adults – about the pandemic and the precautions they might take to avoid infection. Given the median time of ten years between infection with HIV and conversion to AIDS, few high school or university students ever meet anyone in their everyday world with the virus or its later symptoms. The pandemic does not seem immediate and at hand, it appears to be happening somewhere else and to someone else, so "why should I worry?"

This insidious lulling effect works its way up to the highest levels of government to produce a kind of "see no evil, hear no evil," so why bother to speak about it and do something? Early funding for research and caring was slow, and within the huge medical bureaucracies responsible for allocating the available funds there was an extraordinary complacency and lack of urgency in the early years. That, however, is another story. What we can see here is the way in which the long incubation period, and the apparently slow, but essentially unseen transmission of the virus in the early years of the pandemic, create conditions in which it is difficult to generate serious and hard thinking about breaking structures of transmission. So if quarantine is out, what are the alternatives?

Perhaps the most obvious is a technological solution, a search for the technical fix in a technically saturated world. Surely we can come up with a vaccine, or even a cure, something that will knock the HIV out? After all, we have done it before – just think of all the diseases we can protect people against today – so let us pour money into virology and pharmacology, and see what science can do. We now realize that this solution – vaccine or cure – looks very far off. The HIV is enormously complex, it can hide and so protect itself in a number of different cells of the human body, and like all viruses it has a propensity to mutate. Even after the first two years of AZT treatments to slow down the collapse of the immune system, doctors in San Francisco were already seeing the emergence of AZT-resistant strains. Highly experimental vaccines produced at the end of the first decade from HIV strains available in the early and mid-eighties were useless against later mutations, and a US-HIV and Thai-HIV have been recognized as significantly different and needing different vaccines. In any case, and no matter how much money we throw at it, testing the efficacy of a vaccine takes time, and that time is almost impossible to shorten in any significant way. Breaking transmitting structures by making those elements-at-risk in the set disappear looks like a long, difficult, and perhaps impossible, task.

Moreover, our thinking, even our geographic thinking, is still confined to the individual level, where admittedly everything ultimately begins and tragically ends. But the fragmenting of backcloths can take place at very different geographic scales, starting with aggregations of individuals in relatively small areas that circumscribe people with particular social, cultural and sexual characteristics. The Castro District in San Francisco, for example, is a haven for homosexuals, and notice how even the word "haven" (harbor) implies a sheltering and degree of disconnection from rougher social waters. Similarly, a crack house in the Bronx will draw clientele from a relatively limited area around it, with a high rate of intravenous drug and heterosexual transmission in the group of people involved. Equally, a small area in a city noted for people with devout religious observances (perhaps Hasidic Jews), or a rural valley peopled by the Amish of Pennsylvania, will be somewhat isolated from more worldly connections and so be partially protected from infection.

In none of these geographic aggregations, however, is disconnection complete. In the Bronx (chapter 10), you can see distinct distance decay effects – declining intensities of contacts – around major crack houses, but these still form potent sources of infection along chains of individuals to the "outside world." Similarly, the Castro District is hardly hermetically sealed, either from a larger homosexual community that ultimately connects worldwide, or from the much more local heterosexual communities surrounding it. In the largely heterosexual districts adjacent to the Castro, 2 percent of the women regularly have unprotected sexual relations with homosexually-active male partners, and this estimate is likely to be low because the interview conditions on which it is based guaranteed confidentiality but, by definition, not anonymity.

As we go up the geographic scale to larger and larger areas we begin to aggregate more and more people of very different characteristics. But even at these much more general geographic scales we can see connections along which the HIV is channeled, although political decisions may attempt to disconnect, isolate and protect against transmission. At the national level, the United States officially forbids entry of HIV infected people, although it makes no attempt to test anyone except those seeking permanent residence, while the former Soviet Union required HIV testing of all students and visitors staying more than a few weeks, and instantly repatriated any foreigner found seropositive. Even though there is an emerging relationship between international connections and HIV infections, few countries are really isolated in today's world, although after the first decade Bhutan, North Korea, Albania and Mongolia had yet to report any cases. Already the pandemic is leaking rapidly from Thailand across the borders with neighboring Burma, China, Laos, Cambodia, and Malaysia (chapter 8); and with the more open frontiers (and so greater connectivity) between the countries of Eastern and Western Europe, the virus will be transmitted much more rapidly over the restructured national backcloth.

The diffusion of the pandemic to Eastern Europe will be particularly tragic because the relative isolation of these countries until quite recently meant that little HIV traffic had been transmitted. In a sense, countries like Poland, Hungary and Czechoslovakia are where Switzerland, France and Spain (with the highest rates in Europe) were ten years before. The early eighties were a time of complacency when the pandemic seemed barely a remote possibility on the far horizon, but in fact it was precisely the time when intensive efforts at widespread and open education should have begun, and the protective means made available to prevent transmission.

What are these means? They are the availability and cultural acceptance of the latex condom, something eastern Europe is unlikely to have readily available on economic grounds unless priorities are radically and swiftly altered, and unacceptable for many on religious grounds in areas where the Catholic church has great and persuasive force. Yet the condom is the only protection we have, the only breaker of those intimate and very human relations that structure people into a backcloth for HIV transmission. Condoms are structure breakers. With no vaccine, and no cure, education is all we have. I simply do not take seriously religious preaching of any sort as an effective way of preventing many young people dying at a tragically early age from an infectious disease that could have been prevented. Preaching total sexual abstinence to young people in the first and emotionally intense throes of adolescent sexuality is preaching a prescription for death. Further arguments for contraception – the elimination of unwanted pregnancies, unwanted children, and often illegal, and always traumatic, abortions – I simply leave aside as obvious to any thinking and feeling person. In a world with an HIV pandemic, the latex condom emerges as the only practical and responsible strategy. And as the only meaningful response, we shall have to examine its geography.

5

Transmission break: a geography of the condom

When you distribute condoms as a way of not dying, it may be a different ballgame.

Robin Ryder, Project SIDA, Mama Yemo Hospital, Zaire

I'm not afraid of dying or anything else . . . I wasn't born to live forever and I have to eat.

Luis, 14-year old boy prostitute, Rio de Janeiro

. . . on behalf of the child's right not to lose his or her life, I vote Yes.

Westina Mathews, voting the deciding vote to distribute condoms to New York City teenagers

I suspect that "a geography of the condom" will strike many people as a strange subject, even though they would accept quite readily that the condom has an interesting history, from the fishskins employed by Regency bucks and other men about town across eighteenth-century Europe, to pictures of the British prime minister and the queen (separately, not together) on condom packets in Victorian times. After all, said these most proper portraits, even "nice people" may use condoms. But condom use varies in space as well as time, and it has a complex geographical dynamic that we must try to understand, especially as it is the only structure breaker we have, the only barrier to the spread of HIV. To help us think about the condom's geography, I propose we try a thought experiment, an exercise in both spatial and temporal imagination.

Suppose we had a map of the world, and we wanted to show in general terms the intensity or rate of condom use as a contoured surface, rather like the usual weather maps in the newspaper or on television, with the isolines showing areas of high and low pressure. Only on our "condom surface" the highs would be over parts of the world where condoms were frequently used, while the lowest areas would indicate their virtual absence in human sexual relations. What would the condom surface look like? We can imagine high peaks of relatively intense use over a country like Japan, where the contraceptive pill is still illegal; over the countries of Scandinavia, with their open sexual education; and perhaps a few sharp peaks of use over countries like Singapore and Taiwan. Over the rest of the world the condom surface would be moderate to low. In Europe we would see a steep decline, a sharp condom gradient, from the northern countries to the Mediterranean South, matching at a smaller geographic scale a similar declining gradient from northern to southern Ireland. Both would match closely the changing mix of Protestants and Catholics, although this underlying religious reason is obviously not the only one. In the United States, for example, the surface of condom use would still remain at moderate levels even in areas where the Catholic church is well represented, simply because most American Catholics choose to ignore a supposedly celibate patriarchy.

Obviously this condom surface is not a static one, but changes over both historical time and at more immediate temporal scales. Historically it would barely be perceptible until the nineteenth century, when the strange elastic properties of "India rubber" and Brazilian latex made modern protective sheathing of the penis possible. Prior to the widespread availability of natural rubber, and the discovery of the vulcanization process, fish and animal skins had been used, and they are still in use today, even though the animal membranes are not barriers to the passage of the HIV, so they cannot serve as the structure-breakers we require. The condom surface has also changed markedly in the last 30 years. With the availability of antibiotics, condom use as a protection against other sexually transmitted diseases, like syphilis and gonorrhea, declined relative to the growing populations that might have been potential users, and their contraceptive use plummeted with the introduction of the pill. All that has been reversed with the HIV pandemic, and condom use, with its concomitant demand for latex, has soared in the last decade. In our thought experiment we can imagine the HIV diffusing on a worldwide scale, changing the condom surface like a cold front rippling across a weather map.

A cold front of HIV is only a very rough analogy in our thought experiment. While the general level of the surface is rising, the marked variations in condom use from place to place and country to country are the result of highly complex and interrelated effects of economic, religious, political, cultural and even aesthetic realms of human life. Condom use mirrors, to a high degree, the great variety of "worlds" in which people around the globe find themselves. A rural African may well be prepared to use a condom, but as one medical worker noted, "He [may be] in a setting . . . where the country's public supply has been out of stock for some months, popping out to the chemist's [pharmacy] involves a trip of several days, and the price of a packet of three could feed a family of ten." On the other hand, a relatively rich and married Brazilian male in Rio may pay up to three times the going price for anal intercourse with an 8-year-old "Prince of the Night" if a condom is not used.

Acknowledging such variations in condom use, and the complex mix of reasons for them, brings us face-to-face with the geographer's claim that *place matters*. No grand theories are going to help us here, not the least because grand theories in the oxymoronic human sciences never predict anything beyond the banal, and their supposedly grounding tautologies are essentially untestable. The geographer acknowledges that it is only in place that we can observe a particular level of condom use, observe how it changes over time, and speculate about the particular mix of underlying reasons for the particular level we see in a particular geographical and historical setting. Any attempt to formalize these place-specific rates of use into a global predictive model assumes stability of human behavior over space and through time, an assumption that sends real geographers and historians into fits of laughter. This makes the human pseudoscientists very cross, whether they are grand theorists extruded from the heydays of neomarxism, or little human engineers trying to fit computed and purely mechanical mathematical functions with high speed machines to predict condom use around the world. In either case we shall leave them to play with their toys, and try here to get a sense of the enormous variety in condom use and some of the many reasons for it.

In Europe there are steep gradients on our condom surface (figure 5.1), from an always relatively high and undulating plateau of use over northern Scandinavia, northern Germany, Holland, Belgium and Britain, down to the Mediterranean of Spain and Italy, and east to Poland, Czechoslovakia, Hungary and other portions of the former Soviet empire. At a general scale, these low plains of condom usage roughly coincide with the influence of the Catholic church, but this is too simple an explanation by itself. The Catholic church rejects condoms outright, and influences other institutions of Eastern Europe to the extent that even the Polish Medical Ethics Commission labelled condom promotion as "shameful." Condom promotion is still forbidden in that country, but this will undoubtedly change as more young people die and as reason overrules belief. On the other hand, the hit song of a well-known pop singer in Poland warned about AIDS and advised

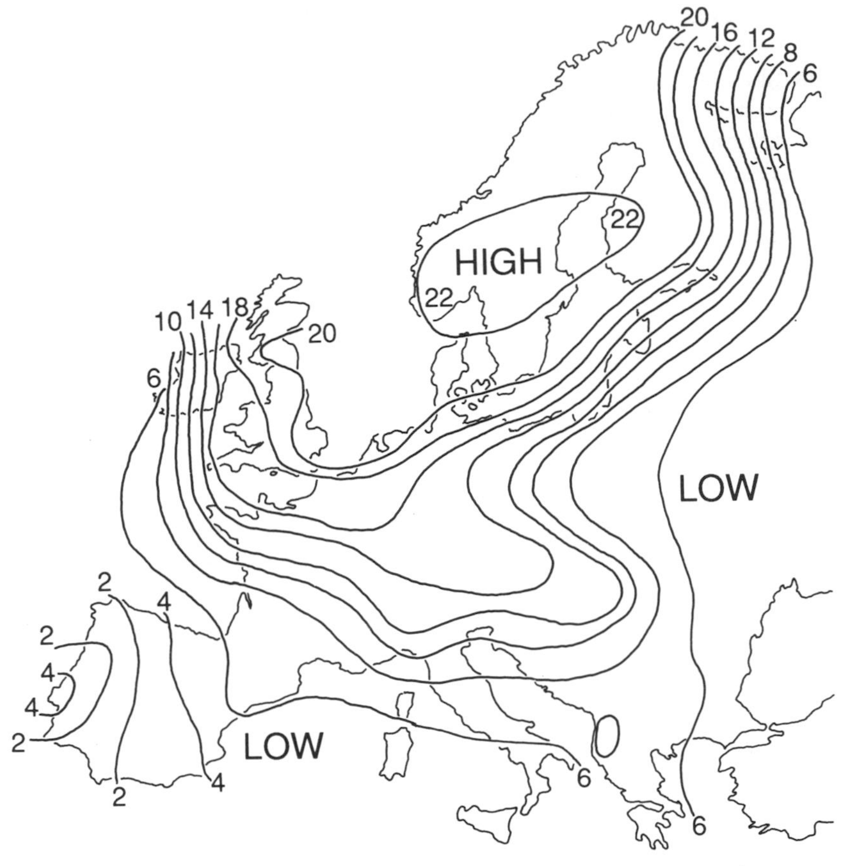

Figure 5.1 A hypothetical, but reasonably plausible "condom surface" over Europe, indicating plateaus and valleys of relatively high and low use. The variations from place to place would be the result of many highly complex and interrelated economic, political, religious and cultural factors. Numerical values indicate relative differences only, but could be scaled appropriately.

condom use as a protection. This is an important educational theme that we shall come across again. But the problem in Poland and the rest of Eastern Europe is primarily whether there are any condoms to buy, and if they are affordable to most people without state subsidy. We are talking about a part of Europe where abortion was often the only means of "contraception" for half a century, and where many women went through ten to fifteen of them during their procreative years. With German help, Hungary is installing condom vending machines, but in Russia, during 1991, an imported condom cost up to $17 on the black market. In this part of the world, religious fervor and economic scarcity work hand in hand to lower the condom surface.

But the situation in Western Europe, where the supply is plentiful and the economic constraint is seldom binding, is equally complex. In the Netherlands, bowls of condoms with a delicate drawing of a bee fertilizing a flower are available free in bars and nightclubs as reminders of the pandemic, but in France the use rate is only 3 to 7 percent, condom advertising was prohibited until 1987, and advertising its contraceptive function was illegal until January, 1991. Most Frenchmen express great distaste for using condoms on sensual and aesthetic grounds, even the very thin and electronically tested varieties imported from Japan. For the French, condoms are "English riding jackets." The English respond by calling them "French letters." France has the third highest rate, and 60 percent more people with AIDS compared to any other country in Europe, with about 20,000 cases as this book goes to press. In fact, the farther south one goes into the Latino cultures, the stronger the machismo, and the lower the condom use. It should come as no surprise that Spain and Italy have the second and fourth highest rates of AIDS in Europe.

A pervasive feeling of male dominance in these societies does not help either. Even female prostitutes are reluctant to insist on condom use since it might imply that they are already infected, or result in their male partner taking offense if his pure state of cleanliness is questioned. We shall meet this same dilemma in Latino cultures on the west bank of the Atlantic River. The infection rates are also increasing steeply in Eire, where the rigid attitudes and political power of the Catholic church have produced a generation of cowed politicians who for many years proved incapable of reversing a legal ban on the sale of condoms. These head-in-the-sand attitudes mean that AIDS education in schools started only after the first decade of the pandemic, and after severe objections and behind the scenes machinations, a delay that resulted in the infection and death of many people in their twenties and early thirties. Homosexuality is still illegal, which means that many men disguise their sexual orientation, and since a high proportion are married, and therefore bisexual, there is a rapid seepage of HIV into the heterosexual community. Eire also has extremely high rates of emigration, with many people returning permanently or on visits, meaning that Eire is highly connected into the international backcloth for HIV transmission.

In North America the HIV "cold front" has had a remarkable effect as heightened awareness has slowly lifted the condom surface. Like the homosexual communities of San Francisco, who saw half their friends convert to AIDS and die, the larger heterosexual community is beginning to realize that the HIV is not something "out there," far away and remote. But the change in protective use is still slow. In Canada, after the first decade of the pandemic, only 19 percent of university students regularly used condoms for fairly casual sexual relations, and at my own university in the United States (Penn State), a very careful longitudinal study showed that only 50 percent of a large group of students used condoms regularly, even though they had been the focus of a special educational campaign, many aspects of which were conducted by the students themselves. Nevertheless, the openness of discussion can only save young lives, and sometimes even an element of relaxed lightheartedness in discussions about condom use can spread the word faster. Two students started to produce custom-made condoms imprinted with mascots in school colors, and increased their sales from $30,000 to $70,000 in one year.

The heightened sense of awareness also reaches down to the high and grade school levels, particularly in major metropolitan areas where infection rates are already measured in percentages (per hundred) rather than the "per thousands" of only a few years ago. Over 10 percent of street youth in New York city are infected, and similar levels are found in the cities along the eastern seaboard all the way south to Miami in Florida. After a long drawn out battle, the New York Public School System began making condoms available to teenagers but, for all the millions spent on official education, the most effective people may well be the rap singers who reach the same audience far more effectively with their lines about "Jimmy Caps." Elsewhere in America the message varies widely: in some places the AIDS pandemic is barely mentioned; in others the grade school children are remarkably knowledgeable. In a local school near me children in a fifth grade class (10-year olds), were discussing the effects of alcohol, and one young man said seriously "Yes, and if you do silly things when you're drunk you could also get AIDS," a remarkably percipient comment in light of the fact that 80 percent of sexual relations among college students involve alcohol.

In North America, most attention during the first decade of the epidemic has focused on the exponentially soaring peaks of infection in the major cities, but the HIV is also spreading to the rural areas, often in association with the use of crack among high school students. Crack has a particularly insidious effect, for not only is it highly addictive, so that young people will barter sex with dealers to get it, but it also produces a drug high with a moderate degree of sexual impotence in the male. As a result, bouts of intercourse are prolonged, and with unprotected sex small tears and lesions provide gateways for the transmission of HIV. Even these may be unnecessary: we know now that certain cells in the mucous membranes (mouth, vagina, colon) can be penetrated by HIV without any physical tear, sore or lesion present. In rural south-eastern Georgia a rapid increase in infection in high school teenagers was attributed to drug dealers from Florida moving out from the I-95 interstate highway, hooking young people on crack, and bringing an HIV wave close behind them.

Obviously our condom surface is the result of many interacting and highly place-specific variables, among which education, and what we might call "class attitudes," may be important. In one brothel in Nevada (a state where they are legal and medically inspected institutions), one owner insisted on compulsory use of condoms, and made it clear that no woman would lose financially if she refused a customer on those grounds. The owner received work applications from prostitutes all over the United States, but he was obliged to show a dozen or more customers to the door each week when they refused to use condoms. One hardhat walked out declaring "I'd rather die than use a rubber!", a wish likely to be granted since prostitution remains a major source of HIV infection.

An overriding sense of male pride, of rampant machismo, also characterizes most of the Latino cultures of Central and South America, and here the low level of our condom surface also reflects high levels of ignorance and poverty, as well as Catholicism often translated into political and legislative influence. In El Salvador, for example, poverty and prostitution go hand in hand, and the low status of women makes it virtually impossible for a woman to refuse unprotected sex if a customer demands it – as he usually does. For the same reasons an infected women will remain silent even when she suspects or knows she is infected for fear of driving the customer, her only source of income, away. Nor is the pandemic slowed down by the common custom of fathers taking boys on their fifteenth birthday to a prostitute for instruction, so inculcating at an early age a feeling that being a man means having as many women as possible. Yet El Salvador, quite typical of many other Central American countries, is projected to have an HIV infection rate of 11 percent by 1996 as a result of high rates of transmission internally and externally. Nearly one fifth of the population have lived in the United States during the war-torn years of civil strife, usually in areas with large numbers of poor Latino people where the pandemic is already well-entrenched through the widespread custom of using prostitutes at weekends.

Further south the high, but still grossly under-reported, rates of HIV and AIDS are almost a mirror image of our low condom surface. In Colombia, condoms are simply not available in sufficient quantities, particularly outside of the major cities like Bogota, Cali, and Medellin, the center of the huge cocaine trade, and machismo anyway ensures that their use is minimal. Brazil has the distinction of leading the pandemic in South America, with infection going hand-in-hand with poverty, particularly in the *favelas*, the huge slum and squatter areas surrounding most of the major cities. In Brazil's prisons, 20 percent of the males, and 38 percent of the females were infected by 1989.

The Caribbean is also a disaster in the making mainly because condom use is treated as an amusing irrelevance by the prevailing cultures. In Jamaica, free condoms in the university go unused with very high rates of casual sex among students, the educated elite of the Caribbean, and startlingly fierce "Panthers" (the local brand name) on one or two billboards around town hardly seem to help. This is a culture with urban and rural pockets of abject poverty, where 26 percent of the working people are unemployed; where large numbers of agricultural workers emigrate to Florida; a country where the average age of a women bearing her first child is 15; where men sometimes try to "score" their half century in children, as though procreation were a cricket match; and where family planning for most people is considered a joke. One university-educated woman with years of experience in social and medical work refused to have anything to do with family planning and contraceptive education. "I see no reason," she said, "to waste my life on something on which I know I cannot have the slightest impact." There are areas of Kingston, Jamaica's capital – containing over half the people in the country – which even armed police are reluctant to visit. And if you live in a squatter area with one standpipe of water for 50 families, and perhaps a small kerosene lamp at night, there is nothing much else to do except make love without a condom you cannot afford, and which you would probably not use even if it were free. It is difficult to think of more tightly connected chains of potential transmission for the HIV traffic. Meanwhile, USAID, not the choicest acronym, plans to make Jamaica an exemplary showcase for HIV intervention, when the Jamaican authorities themselves have little idea where the infected people are, or how to reach them in the sullen and violent slums where many of them live.

Over Asia our condom surface varies greatly from high peaks of use over countries like Japan, to virtually zero levels over Burma, a country where condoms have been outlawed for 30 years or more. Japan has long had high rates of condom use, because its educational system heightens an awareness of the need for birth control in a country with high densities of people in large urbanized areas, and where the contraceptive pill has been outlawed. Infection rates are small but rising, mainly as a result of foreign contacts. In fact, AIDS has been considered a "foreigners disease" not to be caught by sexual relations between the Japanese themselves. Unfortunately, Japanese men are major customers of a huge international sex industry in south-east Asia, spanning an arc from the Philippines to Thailand involving hundreds of thousands of women. High condom use has resulted in remarkably low rates of infection so far, but heterosexual rates of transmission may rise slowly, much of it through prostitution. A shrine at Kanayama now offers small phallic amulets to protect prostitutes against HIV, but their effect has yet to be evaluated by epidemiological studies.

The situation is quite the reverse in India despite a remarkable openness brought about by years of concern for voluntary population control by family planning. Condoms are widely advertised on billboards, but the message emphasized is always contraception not disease control. Infection rates are as high as 40 percent in the huge sex industries of Bombay and other major cities, and many of the women come from as far away as Nepal, assuring widespread transmission of HIV when they return. We shall meet this problem of long distance transmission again when we take a close look at Thailand (chapter 8). But the major problem once again is poverty and ignorance. In many rural areas of India people simply do not believe that the HIV is around, even though thousands of the men in the region may have gone to a big city and returned over the last decade. They either could not afford condoms, or decided they did not like them, and so the HIV comes to the smallest rural village, generally via heterosexual transmission. Housing is crowded, with many children and adults to a room, and sexual relations are often by stealth, making condoms difficult to use, and even more difficult to dispose of. As for use by young married couples, there is such an emphasis on female fertility that any delay of the first child, and even subsequent children, would bring embarrassment, ridicule or scorn to the young husband on the grounds that he was impotent.

In both Indonesia and Sri Lanka, the low level of the condom surface reflects rather similar cultural, political, and economic conditions. In both countries condoms are difficult to obtain, unless you happen to be wealthy and well-educated and can afford foreign imports available in a few large and western-type supermarkets or pharmacies, and any mass media advertising or even mentioning is prohibited. Ironically, both Sri Lanka and Indonesia are exporters of raw latex, but they lack the capacity to turn this product into life-saving condoms. Yet both countries are well-known for their high levels of male prostitution, a knowledge of which Sri Lanka ignores and Indonesia suppresses in order to keep up the tourist trade. Many of the tourists to Sri Lanka are males seeking the services of the 2,500 beach boys south of Colombo, a group already HIV infected. Any male foreigner walking alone around the cities of Sri Lanka will soon be solicited by young men for sex either with themselves or their "sisters." Even well-intentioned efforts to raise condom use in south-east Asia may go awry. In 1973, large shipments of American condoms to Thailand proved less than satisfactory because most of the men were obliged to attach strings which were then tied around the waist to ensure they did not slip off. We have to be sensitive not only to cultural variables but to physiological ones too.

Over Africa, despite real efforts at education and making condoms more widely available, our condom surface is generally very low, and it mirrors the appallingly high rates of HIV infection in this most gravely hit continent. So terrible is the pandemic that we must examine it at a continental scale (chapter 7), but again the variation in the condom surface reflects an extraordinary mix of place-specific cultural, economic, religious and political variables. In Ethiopia, for example, the patriarchs of the Ethiopian orthodox church refused to allow condoms to be mentioned in any AIDS educational literature, and condom use is to all intents unknown among the huge number of bar girl prostitutes that are available in even small towns, particularly those connected by major truck routes. As we shall see, long haul trucking from Somalia to Zambia forms a major connecting relation on the international backcloth of Africa, and in Tanzania the truckers themselves have formed the Truck Driver's AIDS Intervention Project, carrying condoms from truck stop to truck stop, often advertising their product on the sides of the trucks themselves. In the last six months of 1990 they distributed 725,000 condoms free, but this subsidized attempt has to be seen in the light of the 800,000 people presently infected in Tanzania alone.

Matching numbers of condoms to numbers of people is also discouraging in Uganda, where USAID distributed two million, enough for one-third of the couples in the country to have protected intercourse once. In fact, most couples have never used any protection, and even after an extensive educational campaign, 16 percent had still never heard of them. It was only in June, 1990, a decade or more into the pandemic that the President gave his reluctant support in the face of great counter-pressures from the Catholic church. About 50 percent of Uganda's people profess Catholicism, and it was in Kampala that the Pope preached against their use. An advertisement for condoms in the *Kampala Weekly Times* saying "The bible may save your soul, but this [condom] will save your life" was hurriedly withdrawn when Catholic clergy protested.

Next door, in Kenya, education and condom availability are again a matter of too little, too late. During the early- and mid-eighties, politicians dealt with the pandemic by the simple expedient of denying it existed, and condom education, use and availability were grossly delayed. Once again we are talking about a country in which only 26 percent of the rural people have access to running water, and only 6 percent to electricity, and condom disposal, let alone use, is a severe problem. Over 90 percent of the prostitutes of Nairobi and other major towns are HIV infected.

Mozambique faces similar delays and problems. Sexual education is forbidden in the schools, and condoms are essentially unknown and unused in a country where girls will start to marry at the age of 12. It is a country in total political and cultural disarray, where AK-47 assault rifles are readily available to 12-year-olds (who use them to hold up small stores for candy), while the pandemic continues to rage for the lack of *any* of means to stem it.

In Zaïre, a major center of the pandemic, the condom surface has been raised slightly by the HIV passing through, and "Prudence" condoms are advertised widely in the major cities. One pop concert required people to show a Prudence packet as a ticket to the show, and again pop singers appear as the most potent educational force for young and highly vulnerable people. An American organization, US Population Services International, increased its distribution of condoms from 200,000 in 1986 to nine million in 1990, but even these would allow one or two protected episodes for every adult in a country of quite free sexual relations. By 1988, 20 percent of the staff (doctors and nurses) of a hospital in Zaïre were found to be seropositive, not through treating AIDS patients, but by their own unprotected sexual relations. Nurses in Africa have generally been considered safe, since they had ready access to penicillin and other antibiotics if they contracted various sorts of sexually transmitted diseases. Unfortunately, these are useless against the HIV.

Education is always a difficult problem, and those trying to instruct others in proper condom use may not always get their message across as they had intended. At one family planning clinic in Zimbabwe a woman health worker demonstrated a condom's use by carefully opening the packet, squeezing the air from the receptacle tip, and rolling the condom down over her upright thumb in lieu of any other artifical phallus. A few months later she received an outraged protest from a young man who insisted his girlfriend had become pregnant, despite the fact that he wore a condom on his thumb each time they had made love. At Karoi, in northern Zimbabwe, one truck driver noted that he had overcome his dislike of condoms by cutting the tips off.

Cultural proclivities, customs and beliefs make it difficult to raise our surface, even when condoms may be readily available. In Khartoum, Sudan, the strict Islamic prohibitions on sex for unmarried women appear to be adhered to more strictly than those concerning young male boys. Lacking condoms, 7 percent of the 6- to 14-year-old street boys were infected by 1988, and what the bisexual transmission to wives and families is no one knows. Sudan reports one of the highest rates of infection in the Islamic world, but most cases go unrecorded. Cultural beliefs are also a barrier to condom use in South Africa, where many people consider it essential that the sperm of the man actually enters the woman. Condoms prevent this, with the result that there is a high degree of reluctance to use them – as in the case of the Zimbabwean truck driver. In a country with very high rates of labor migration, and prolonged stays in work areas where families are forbidden, the HIV rate is high and the virus is guaranteed to spread quickly by heterosexual means to the home areas.

One part of the world where our condom surface may have risen quite dramatically in the past five years is Australia, perhaps in part because AIDS cases were reported early by public health systems, and partly because condoms are readily available and easily affordable. So sudden was the rise in use that several municipalities in New South Wales found their waste disposal and treatment facilities clogged by condoms, and they had to issue an appeal to dispose of them "more thoughtfully." Condom use has also been greatly helped by the strong union of sex workers, particularly in the state of Victoria. Most union members of the Victoria Prostitutes Collective insist upon their use as part of their educational "Hello, Sailor!" campaign, and as the forthright president of the union noted "The virus doesn't travel on dollar bills."

This, however, is perhaps the one bright spot, one rising and local plateau of condom use on an otherwise depressed surface worldwide. And in this generally depressing geographic picture there are two aspects which call for special comment. The first is that the HIV appears to be spreading where, in general, the status of women is low. Low status may frequently go hand in hand with low economic and educational levels. In Brooklyn, a borough of New York City, there is often no discussion of sex between a latino man and his wife, and few latina women would dare to ask their sexual partners to use a condom, or even broach the subject of HIV testing. One member of the Brooklyn AIDS Task Force noted, "For some of the poorest latina women, sex is something [that] happens to them in the dark and in silence, something they do not talk about even with their husbands." As we saw before, the latino culture, whether specifically from Puerto Rico or throughout Latin America, also prizes virginity highly at marriage, leading to higher rates of anal intercourse in young people, virtually all of it unprotected, but employing a sexual relation that is much more likely to result in HIV transmission.

Such a subservient status of women is almost universal. In Uganda, 30 percent of the women felt that they had no control whatsoever over their sexual status, and we are talking about a country of frequent road blocks by the military who will often make sexual intercourse with a woman the price for letting her continue her journey. In Tanzania, women say that insisting on condom use will jeopardize their relationship with a man, and in the society as a whole a woman has virtually no legal rights either in marriage or upon the death of her husband. It is the same in India with the balance of power in a marriage always heavily in the man's favor. And in Ecuador and Malawi and Indonesia and . . . Japan and Spain and Yugoslavia . . . and . . . so half the human family is held in a subservient relation with the other half, and the HIV spreads farther and farther under this almost universal and unequal power relationship.

The second, thoroughly depressing phenomenon affecting local levels of condom use is that many members of the homosexual communities that survived the trauma of the eighties are reverting to unprotected sex, and their behavior is being followed by a new generation of young men who have adopted what appears to be a death wish. From London to San Francisco to Santiago, Chile, to Auckland, New Zealand, to Australia, the same message of despair, voiced by the men themselves, is coming through: we are despised, we care little for ourselves, we have no sense of self-worth, we are people facing death with no openness towards procreation and life. So what the hell, what's the use? In London, many of the young men at homosexual bars like The London Apprentice scorn the use of condoms, often voicing their objections in streams of bravado and fatalism. In San Francisco even some of those who saw half their friends convert to AIDS and die are reverting to unprotected sex, and 70 percent of Australian homosexuals have reverted to non-condom sex all or part of the time. From Santiago and Auckland the message is the same: we are outcasts, pariahs, nobody cares, why should we? Eat, drink, be merry, have sex, and die. This unprotected episode will be my last. At least until the next time. Small local peaks on the condom surface, peaks over the most highly infected groups in the world, have collapsed. And they will subside further in the years ahead. If people really wish to die, no one can stop them.

Whatever the geographic scale at which we choose to think about the spread of the HIV – from town to town, city to city, region to region, country to country – we know that the virus moves at the most local and personal scale from person to person. This is the ultimate backcloth of connected people on which the virus travels, although we can never actually observe it directly. The reason is quite simple: those relations structuring the set of people are usually sexual relations, some of the most intimate, the most private relations we have as human beings.

Only the condom can break those connections over which the HIV moves, and yet still allow people to love each other physically, to exercise their sometimes mutual lust, and to share some of the most intense and driving physical moments of exaltation known to us. How many lives will be saved by condom use no one knows. As we have seen, many things work singly or in tandem to reduce the height of that condom surface around the world, those contoured hills and valleys that portray the geography of thin tissues of latex that can protect and save people's lives.

So the HIV travels on a non-observable structure of human relations. Only after this hideous traffic has been transmitted can we observe at some larger geographic scale its deadly effects. In a sense, we always have to raise our geographic sights, from the individual to groups of people in certain places at certain times. As geographers we have to aggregate, and in the process we can be drawn away from individual caring to that reified world of aggregate "things." There is no help for it; to understand this pandemic we must be prepared to think about it at all geographic scales. But to do this properly, and before we examine actual examples, we shall have to step back and think about how things spread over geographic space and through historical time in more general ways, in terms of the underlying principles involved.

6

How things spread: hierarchical jumps and spatial contagion

Whilst my physicians by their love are grown Cosmographers, and I their map, who lie flat on this bed.
John Donne, Devotions Upon Emergent Occasions, *1624*

For a geographer, a fascination for things spreading over space and through time, no matter what they are – viruses, innovations, fashions, fads, etc. – seems not only natural but inevitable, almost what it means to be called to the geographic way of seeing the world. There is nothing so intriguing, so generative of the "why" and "how" questions, as a sequence of maps showing how something diffuses, whether it is an innovation like the printing press, the first performance of a musical work, an agricultural innovation like hybrid rice, or the annual influenza epidemic. As we shall see, many of these map sequences seem to have a logic of their own: the changes over geographic space at certain places, the appearances of things at certain historical times, seldom jump all over the map in an apparently random way, but often seem to unfold like a photographic plate developing in the darkroom tray. One can often see intuitively how the latest developments, the latest image on the map, was somehow there, latent, in the earlier maps of the sequence.

Given such fascinating cartographic expressions of spatio-temporal processes (to use a bit of jargon), it is little wonder that geographers have tried to find the "rules of the game" underlying the unfolding patterns on the map, not simply to test these by trying to simulate and reproduce the map sequences, but to run the game ahead of where we are at the moment, and so predict what may be coming to us out of the future. The very first computer simulation models in the human sciences were constructed by geographers studying diffusion processes, in the days when computers almost seemed to be made of vacuum tubes connected by hairpins, bits of string and sealing wax. In fact, the earliest computer modeling of geographic diffusion was done in Sweden on a hand-built computer at the University of Lund in the early 1950s. The computer has revolutionized many aspects of geographic analysis, and today only the largest and fastest Crays can keep up with modeling these complex processes. We shall take a closer look at some of these later (chapter 14), and see why it may be important to predict the next map or maps.

But no matter how large or fast the computers, certain basic geographic principles appear again and again, and in one form or another they underlie the most complex and sophisticated approaches. Simply as a matter of intellectual convenience, geographers divide processes of geographic diffusion into two main types, recognizing that in any actual process – for example, the spread of the HIV – both are probably involved in complex and often mutually reinforcing ways. The first is *spatially contagious diffusion*, where, for example, a disease appears to move over the map from an initial source of infection, perhaps a large city or regional epicenter. If we are prepared to mix our similes a bit, it is like dropping a stone into a smooth millpond on whose surface a map has been projected. The regional epicenter is where the stone lands, and we can watch the ripples move out from it, like a wine stain on a tablecloth. In fact the wine stain is probably a better simile, because the wine will stain by capillary action only those fibers in the cloth that are adjacent to those already wet. As far as the spread of a disease is concerned, how far away you are from an epicenter will determine to a very high degree when the epidemic arrives.

This spatially contagious effect was very prominent in past times when distance exerted a strong influence on how quickly people could travel, as villagers well knew when the news of the Black Death arrived by horseman, and they tried to use the early warning to erect barricades to protect themselves from the slower moving but potentially infected wave of travelers. Geographers actually speak of the "friction of distance," realizing that it can change from one region and culture to another, and over historical time. In the olden days, geographic space had a high "friction," and it was experienced as a sticky, molasses-like surface. It was difficult, costly, and time-consuming to move when the friction of distance was high, although it also meant that distance had some buffering and protective effect. Today we live in a slippery world, where distance has much less effect. Everything, including diseases, moves quickly when the friction of distance is low.

Of course, whether we think of stones being dropped into a pond, wine stains creeping over a table cloth, or heat diffusing through a piece of copper sheet, we realize that these are just simplifying images. No map, no geographic space, is cut out of a uniform piece of copper sheet, with the heat from a Bunsen burner under a regional epicenter diffusing by conduction across the map according to the laws of physics. No map of the human condition is smooth and uniform, because human beings themselves structure the space in highly complex ways. They build roads and canals, they use natural or improved riverways to move faster in some directions than others, they build railways and super-highways, and finally connect cities to cities directly with airlines. The actual space through which people move, as well as the HIV riding along with them, is a strangely contorted space. Which is precisely why geographers think about the second major form of geographical spreading – *hierarchical diffusion.*

Hierarchical diffusion means we recognize that things, diseases included, appear to jump from place to place, usually city to city and town to town, without necessarily touching down at places in between. It as though the geographic space on the map were highly structured by strong relations between places in the urban hierarchy, so that what we often observe is a disease or innovation trickling down from the big places to the smaller ones. In most cases it is not quite that simple, and we cannot predict the course of an epidemic just by rank ordering the cities and towns by their sizes, and then finding we have the right sequence of times of first infection. All the same, this simple view is not totally naive. Even if a disease starts in a small town, it is very unlikely that it will be carried to another small town far away. It can, of course, happen: someone picks up cholera in Peru, flies back to Liberal, Kansas, infects two people there by touching food, and then flies to Dubois, Pennsylvania, to join a family for Thanksgiving dinner that turns out to be not a very happy one. It *can* happen, but it is extremely unlikely. What is much more likely is that the infected person will visit a large city, infect other people there, some of whom who will visit New York or Los Angeles, and so start the generally downward trickle of infection through the urban hierarchy.

At this moment we are in what a geographer calls a "gravity model" frame of thinking – a rather technical area of considerable computational sophistication today, but like many ideas that may be difficult to make operational mathematically it is conceptually quite simple. All we have to do is recognize that big cities are more likely to interact by exchanging people than small cities, and places far away are less likely to have many people traveling between them than places near one another. That is why I said we were in an area of gravity model thinking, because human interaction (say, people carrying the HIV) is going to be roughly proportional to the sizes of the cities, and inversely proportional to how far they are apart. We could write $I = P \cdot P/D$, where the Ps are the populations of the cities, say, New York and Chicago or Paris and Lyon, and the D is the distance between them, say 935 miles or 384 kilometers, respectively. I do not want to throw formulas at you, because if they mean anything at all the meaning can only come to us in words anyway, and if they mean nothing in words we should toss them out of the window. But I wanted you to see why geographers call hierarchical diffusion "gravity model thinking": that $P \cdot P/D$ reminds us of Newton's old formula saying that the force of gravitational attraction between two bodies, say the earth and the moon, is proportional to their masses and inversely proportional to the square of the distance between them, or $F = M \cdot M/D^2$, give or take a few constants between not too nitpicky friends. In a geographical context we could put a little α (alpha) as an exponent of distance, say D^α instead of Newton's D^2, and try to estimate what it might be in any particular problem. We can think of it as that "friction of distance," that effect making a map into a surface of cold molasses when it is high, and into a slippery hot griddle when it is low. It does raise some really nasty problems of calibration and interpretation in any actual application, but we do not have to bother with these literally esoteric problems that keep only geographers awake at night.

What does hierarchical diffusion, as seen through this gravity model perspective, actually imply? Something quite simple, but also something quite profound as far as thinking about the HIV and AIDS pandemic is concerned. We must be prepared to shift our thinking from the conventional map of geographic space to a radically transformed and usually multidimensional space of human interactions and contacts that I think we can legitimately call in this book an "AIDS space." It is not such a big and radical jump in thinking if we are prepared to follow the logic of spatially contagious and hierarchical diffusion through.

Suppose we take Ohio (figure 6.1) as a concrete example, and focus for a moment on the three largest cities, Cleveland on Lake Erie to the north-east, Columbus the capital in the center, and Cincinnati in the south-west corner. They have large populations (the first two about 600,000 each, and 400,000 for Cincinnati), and are roughly the same distance (180–225 kilometers) apart. If we think about these populations and distances in our gravity model terms, we are obviously going to predict very high interactions between them, and that translates directly into many contacts involving HIV transmission. And we must not forget that all three cities are highly connected into the much more extensive urban hierarchy of the United States (direct flights to New York, Chicago, Los Angeles, etc.), which in turn is a big piece of the international structure (direct flights to Paris, London, Tokyo, Hong

Figure 6.1 Ohio, with its three largest cities, and the population centers of its 84 counties. Theoretically, the populations of the counties can interact in 3,486 ways, including transmitting HIV. The potential interactions might be in proportion to the number of people in each county and the distance they are apart.

Kong, and so on). We have snipped Ohio out of the larger picture just for our example here, although this does demonstrate how difficult it is to get "closure," to find a piece of our modern world that might be treated for analytical purposes as a closed system. But back to Ohio . . .

Rather than just taking the three biggest cities, we could take the populations of all the 84 counties (or whatever geographical units were convenient and available to us), and then measure the distance from each one to all the others, and work out in our gravity model way all the 3,486 potential interactions between them. It would be a long and boring job by hand, but something a computer could do in the blink of an eye. It would mean that two, rather isolated rural counties far apart on the conventional map would interact very little, which translates into an extremely small probability that the HIV would be transmitted in this way on our Ohio county backcloth. Conversely, and as we have already seen, there would be a very high probability of the HIV jumping from one big city to another but passing over some rural counties in between with few people living in them. The rural counties most vulnerable to HIV infection would be close to and surrounding the large cities, which might quickly become infected by hierarchical diffusion and then serve as regional epicenters for spatially contagious diffusion into what geographers call their *umlands*, the areas surrounding and focusing upon them.

It is exactly these sorts of intuitive but easily formalized ideas that allow us to transform the map of Ohio out of conventional geographic space and into our multidimensional AIDS space (figure 6.2). All we have to do is make the strong interaction or high probability of HIV transmission inversely proportional to the distances our counties are apart in the new and transformed space.

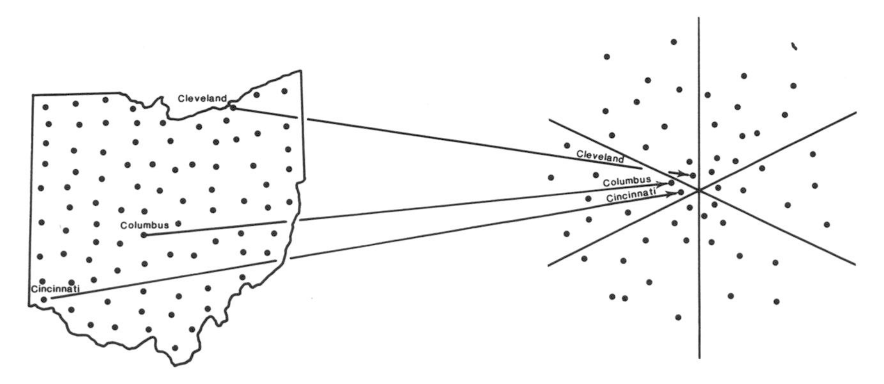

Figure 6.2 Transforming the map of Ohio from conventional geographic space into a multidimensional "AIDS space" with a "gravity model" mapping. Large cities like Cleveland, Columbus and Cincinnati, scattered widely over the usual map, tend to concentrate at the center of the AIDS space due to their high potential interactions which increase the chance of HIV transmission between them.

In this way Cleveland, Columbus and Cincinnati end up close together in the center of the space (limited here to the three dimensions we can show by perspective drawing), while the remote rural counties with few people end up as points at the edge of a multidimensional cloud, literally far from the center and very much on the periphery. Rural counties close to a big city in the space of a conventional map would probably be in the same sector of our multidimensional cloud, although they might also be drawn close to another big city not too far away. Remember gravitation? In our AIDS space, counties gravitate to the "big boys" . . . or should it be big persons? Interestingly, a mathematician would say that in transforming our problem from one space to another we have undertaken a *mapping*.

Imagine yourself now in the center of AIDS space blowing a big soap bubble around Cleveland, the biggest of Ohio's cities. As your bubble expands it will capture counties nearby containing Columbus, Cincinnati, and other big metropolitan areas. As a city–county is captured, it too starts to have a bubble blown around it, and these expanding bubbles capture the counties close to them in the transformed space. Perhaps the HIV spreads in this way, slowly capturing the counties in AIDS space until the multiple bubbles reach those on the periphery and the infection is everywhere. Does the HIV really spread in this apparently simplified way? Recalling that it spreads at the finest geographic scale from person to person, and that we are looking at the diffusion of the virus as it appears at a fairly aggregate scale, the truthful answer is that we can never really tell, even if we were able to gain access to the individual medical files of patients with AIDS. Of course, this would break patient confidentiality (chapter 13), something that is as abhorrent to me as I am sure it is to you, but the purely private information about sexual contacts, IV drugs, or blood transfusions would be useless anyway. No one can tell exactly when the HIV passed from one person to another, and no one – neither the person faced with infection nor the scientific researcher – cares. It is too late.

Nevertheless, and thinking about these cautions, the actual spread of AIDS in Ohio does illustrate the basic principles involved, and since conversion to AIDS follows on the average a decade after the first infection, we are looking at a sequence that mirrors to a very high degree the broad outlines of HIV diffusion roughly ten years before. Geographers know from many examples around the world that broad but invisible channels of diffusion may be extraordinarily stable, often changing only slightly over decades or generations.

For example (figure 6.3), Cleveland already had people converting to AIDS by 1982, and even at this very earliest stage we can see evidence of hierarchical jumps to Akron, Canton, and especially Columbus, and some slight seepage by spatially contagious diffusion around Cleveland itself. The next year Cincinnati starts to become another regional epicenter as the virus (indicated by people with AIDS), jumps to the south-west corner by hierarchical diffusion. By 1984, it has already started to spread like our wine stain into the surrounding counties, and what will soon become a distinct north-east/south-west alignment starts to appear as the Cleveland, Columbus, Dayton and Cincinnati "stains" begin to reach out and coalesce. Notice that the contour lines are not going up in regular (arithmetic) steps, but are increasing geometrically. Like many human things of interest to geographers, the AIDS pandemic produces very tall and sharp peaks on our contour map, and this is the only way we can reasonably show them in the conventional way.

Jumping to 1987, we can see the outlines of two main alignments, the north-east/south-west one which has strengthened in the three year interval, and an east-west alignment across the northern part of the

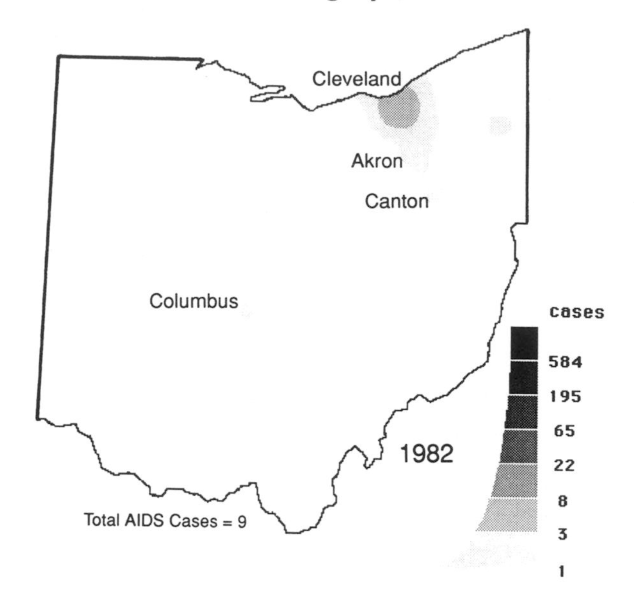

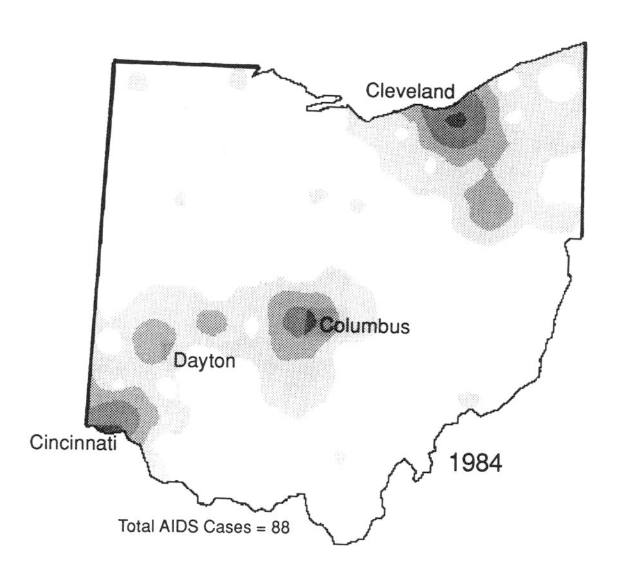

Figure 6.3 The spread of AIDS in Ohio, 1982–90, indicating both hierarchical and spatially contagious effects. A decade earlier the HIV must have moved in a somewhat similar fashion, jumping from city to city, each one of which became a regional epicenter.

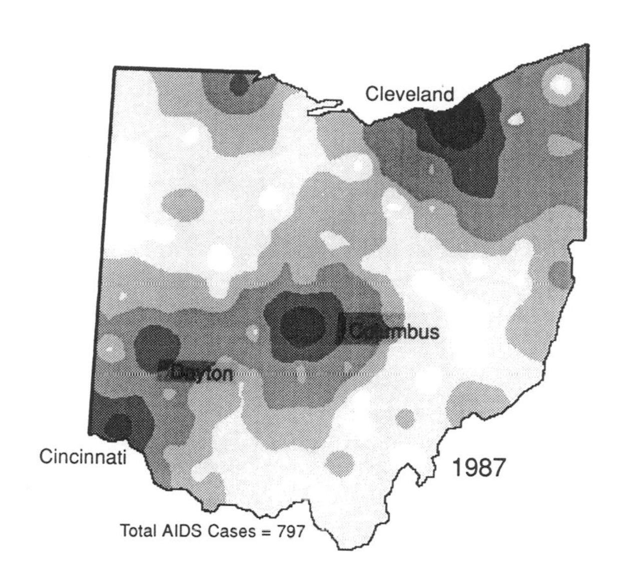
Cleveland
Columbus
Dayton
Cincinnati
1987
Total AIDS Cases = 797

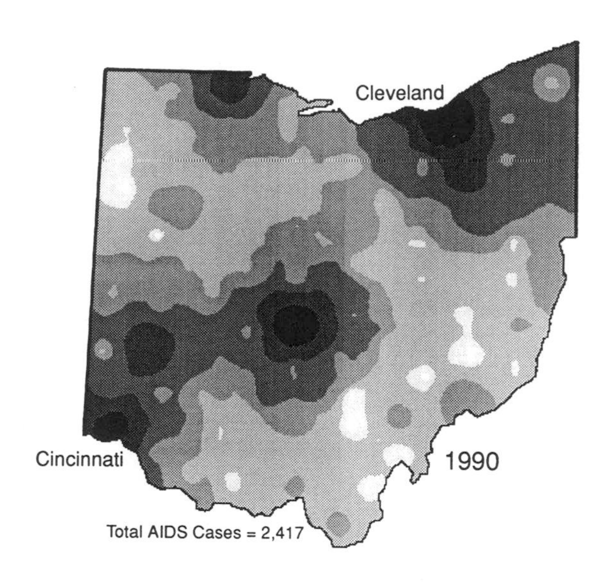
Cleveland
Cincinnati
1990
Total AIDS Cases = 2,417

state. Both of these coincide with major route alignments (I-71 and I-80 respectively), and it is almost as though the transport system is roughly channeling the pandemic at this scale. This hardly comes as a surprise: roads, railways and airlines are built and aligned to serve major flows and interactions of people, and it is people who carry HIV, not wind and mosquitoes. We can be thankful for that. The AIDS pandemic is moving over the space of Ohio and through time not like our heat diffusing across that homogenous copper sheet, but in a geographic space highly structured by the human presence itself.

By 1990, the unfolding pattern has strengthened still further, and the pockets of completely uninfected areas have almost been squeezed out. I think you will agree that we can speak, in some carefully defined sense, about a "spatial logic" in this truly terrible sequence, something we shall take up again a bit more formally (chapter 14).

So our basic principles of spatially contagious and hierarchical diffusion – those oozings and jumps across the map – seem to capture a major part of what is going on at a reasonably large geographic scale. These ideas are also useful handles to hang onto when we go up the geographic scale to whole countries and continents, and even to the global level. We actually met the principle of hierarchical diffusion at the intercontinental scale when we were concerned with the origins of the HIV (chapter 2), as the viral traffic was carried around the world on a network of airlines connecting 51 large cities. Once entrenched in a regional epicenter, and spreading within the urban population, the HIV also starts to seep outwards and across the map in a spatially contagious way. We shall see these two processes at work in the most catastrophic way in Africa, the origin of the HIV viruses, and the first continent to bear the brunt of this terrible pandemic.

7

Africa: a continent in catastrophe

O le SIDA est une terrible maladie
Une maladie qui ne pardonne pas . . .
Luambo Makini, OK Jazz

The donation of USAID of two million condoms to Uganda in 1988 was not enough for even a third of the population to swing into action once.
Dr Samuel Okware, former head of Uganda's AIDS Control Programme

Nowhere in the world can we see the tragic effects of hierarchical and spatially contagious diffusion so devastatingly at work as in the continent of Africa. It is the ur-continent for the entire human family, our common place of origin 100,000 years ago, our genesis as homo sapiens, and the origin of our thousand tongues. There is something perverse in that the cradle of east and central Africa that gave us all birth should now be the origin of a virus bringing us death. Sometimes the gods are not kind. From that nurturing origin it took us, in all our changing human variety, tens of thousands of years to cover the globe, to reach the tip of Tierra del Fuego, the cold wastes of Greenland, the coral strands of the South Pacific. It took ten years for the HIVs, both 1 and 2, to hopscotch around the same globe and entrench themselves in almost every people on earth. The virus is now in native Canadians in the far north, in pygmy people deep in equatorial forests, in Polynesia on once-remote and seldom-visited tropical islands, and in the Shan people of the remote hill country of eastern Burma. In long chains of human relationships and movements every place seems connected to all the others. There's no hiding place down here.

We have already looked at the African origins of the viruses, the crossing over of the HIV-1 and HIV-2 from simian forms in chimps and mangabeys in the east-central and western parts of the continent respectively. Precisely when this occurred we shall never know, but probably within the last 100, and more likely within the past 50 years the viruses moved out of their simian hosts and into people. Whatever the precise moment and period of first faltering propagation, it came at a time when the human pulse of the continent was quickening. Roads were following and radically upgrading ancient paths and caravan trails, new ones were being hacked out of thorn savannahs and tropical forests. Invariably they linked town to town, and then city to city, a lacework of laterite and tar joining up the emergent urban nodes that shine as points of light on all our satellite images. And along that increasingly dense lacework the people moved, from countryside to rural town, from town to major and exploding city. More and more the protective barrier of distance was diminished, and then for the truly wealthy and political elite it virtually disappeared altogether. In the same 50 years the airplane superimposed its own lacework of connections in the skies of Africa and changed weeks apart to hours apart. And along all these distance-breaking networks the HIVs moved with the huge tides of ordinary migrant people, rural to urban, countryside to mine, and with the class elites from political conference to political conference. We cannot examine all of these, and in any case the detailed narrative would become numbingly repetitive, but I want to pick out two exemplifying features: a transcontinental alignment, and a transcontinental focus, two large-scale geographic features that form parts of the complex composite of the pandemic in Africa today.

What is the general geographic picture of the pandemic at this continental scale? We can only compile a map of relative intensities and use it with considerable caution. Official figures reported to the WHO by individual countries are essentially useless: as a UN agency, the WHO is obliged to accept and report them, but we are not. We are not part of the international bureaucracy, and we are under no obligation to play along with officially sponsored ignorance or blatant political lying. At one, anthropologically detached level they are both understandable: at no level are they morally defensible or excusable in the midst of a mortally dangerous pandemic. The politicians in a number of African countries deliberately under-report the degree of the severity of the pandemic out of a sense of shame that such a disease is a sign of backwardness (for years cholera had been unreported or denied outright for a similar reason), or for reasons of economic greed to keep the hard currencies of tourists flowing, many of which find their way back to private and unnumbered bank accounts in Switzerland. We are talking about a continent where the former and self-crowned emperor Bokassa put away $2 billion, Mobutu of Zaïre bled his country of $5 billion, where Ghanaian military officers and politicians simply switch bank account numbers in Switzerland depending upon who is in or out that year, and where Nigerian newspapers report kickbacks of $160 million in two years to a road contractor. Public corruption on this scale has never been seen before in human history, and downplaying AIDS figures is generally considered an irrelevant bagatelle.

In Kenya, the President spent the first four years of the growing pandemic denying that there was any HIV around, and ascribing reports of it to a deliberate hate campaign against his country. He threatened to remove the visa and deport any foreign journalist reporting AIDS, and waited until 1986 before allowing the most innocuous "AIDS guidelines" to be published, meanwhile instructing the Ministry of Health to under-report grossly the known cases on the grounds that many of those with AIDS were "not Kenyans." In the meantime, seropositivity in Nairobi's large prostitute population went from 17 percent to almost 100 percent. What the rate was in their customers nobody knows. In Kenya only the women are forcibly tested, and then criminally charged under Kenyan Public Health Act, Section 17, if they have *any* sexually transmitted disease. In Zimbabwe, a brigader general was appointed Minister of Health in 1988, and promptly reduced the official death toll from AIDS from 380 to 119, and forbade all references to AIDS on death certificates. In this way the AIDS pandemic is officially abolished since there is no official evidence of it. All blood banks with any testing capabilities at all are forbidden to release HIV rates on the grounds that these are state secrets. One doctor in the forefront of AIDS care in Uganda was expelled for discovering and responsibly reporting that over 30 percent of the women coming for pre-natal care to the Kampala hospital were infected. Only in November, 1990, did the President of Uganda permit official endorsement of condoms, by which time more than one million Ugandans were infected. Many doctors with first-hand experience across the continent estimate that actual AIDS cases are 80–90 percent under-reported, and HIV infection rates are largely unknown except for some catastrophically high rates in high-risk groups. In the African context this does not imply homosexuals or IV drug users, but a modernized urban elite with numerous sexual relationships that are considered the norm. The human backcloth in all African cities is very tightly connected, and with a continuous movement of people from these regional epicenters to their countryside origins the HIV is leaking away from these intensely infected nodes very quickly.

In the face of these and many other examples of lying supported by genuine ignorance, the generalized map (figure 7.1) can only paint the geographic picture with a broad brush. Yet even at this scale we can see the way the African continent has been structured by the urban and infrastructural developments of the past half century. Virtually every

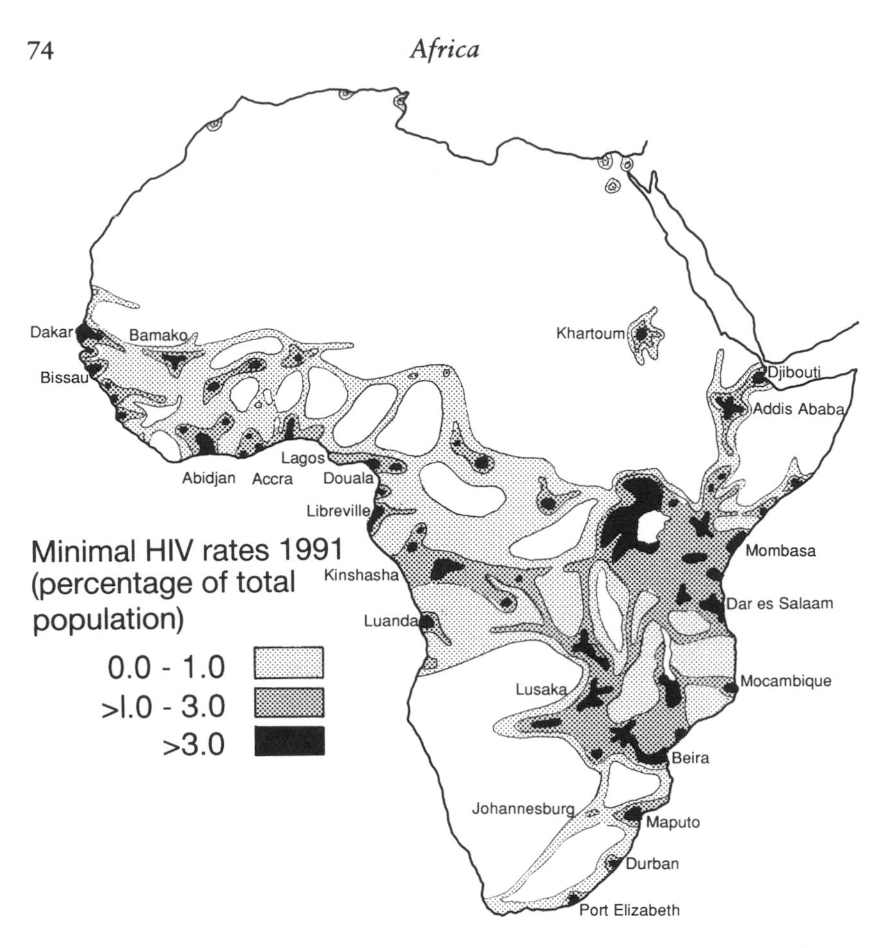

Figure 7.1 The generalized map of HIV infection in Africa around 1991 compiled from many specific and detailed sources. Estimates and extrapolations over space and time from "spot heights" are conservative. In some urban areas rates of HIV infection in certain groups may exceed 90 percent.

large city is an exponential peak infected by the long distance movements of hierarchical diffusion by truck or airplane. The pandemic is especially severe in the cities of East and Central Africa and their rural umlands, as well as in the highly connected cities of West Africa. Only Nigeria, lying between the advancing fronts of HIV-1 from the east, and HIV-2 from the west, is still at a relatively low level, but the virus has reached Lagos along the coast road from Accra in Ghana, Calabar from infected Cameroon to the east, and has arrived in the north via a "right hook" of long-distance trucking to Maiduguri. The same effect of distance from the major continental centers of HIV infection also protected South Africa somewhat during the first decade, although the eventual penetration is shown by a clear north-east/south-west gradient of infection among African women from 1.61 percent in Natal, 0.53 percent in Transvaal, and 0.16 in the Cape. But the distance effect can only delay, not protect. Already the virus is sweeping in from Mozambique and along the truck routes from Zimbabwe, Zambia and Malawi, all highly infected and all traditional sources of mine workers in the Rand. These major routes of transmission form the southern end of the most distinctive alignment of infection in the entire continent, and one we should examine more closely.

The long north–south alignment of HIV in East Africa starts at Djibouti at the mouth of the Red Sea, a port city and terminal for the railway to Addis Ababa, the capital of Ethiopia. Ships entering and leaving the harbor carry more than just cargo: the HIV is not fussy about its mode of transport. At the end of the first decade of the epidemic (1991, and our reference date), over 50–60 percent of the prostitutes, and 1 percent of the general population were HIV-1 infected. Given doubling times measured in months at the beginning of the pandemic, these are conservative extrapolations from the official figures; but here in Djibouti, and as we travel south to Mozambique, there is no reason why we should throw away our commonsense. We know that rates for women and for men are roughly equal in Africa, and as the forthright president of the Victoria Prostitutes Collective noted in a different geographical context (chapter 5), "The virus doesn't travel on dollar bills." In the virtual absence of condoms, rates may be five to ten times higher. A dock worker visits a prostitute on the way home, infects his wife and girlfriends in turn, who spread the virus via other, nearly always multiple liaisons. The "high spots" on the map are estimates of any rate over 3 percent, but we should be aware that in so-called high risk groups – a term that may eventually define Africa – the actual rates may be close to 100 percent.

And so the virus travels up the railway and along the parallel trucking route to Addis Ababa. We shall talk about it spreading in this way, simply because we are making our discursive journey southwards, but in reality every road, north, south, east or west, is always a two-way highway carrying the virus. The truckers themselves are a major group of carriers, and every truck stop along the long-distance routes is a boiling pot of infection, with 80 percent or more of the bar girls seropositive to HIV-1. From Uganda to Mozambique, samples of truckers tested indicated that 30–80 percent are infected, and if we could animate our map for television, say from 1981 to 1991, we would see the viruses diffusing and increasing along the major trucking arteries, with the truckstops and major towns exploding as exponential peaks of infection, forming regional epicenters all up and down the often embryonic urban hierarchy. At Addis Ababa, with casual sexual relations a way of life, convenient small back rooms in almost every bar, and condoms condemned by the church, the official seroprevalence rate in the general population is about 3 or 4 percent. I leave it to you to multiply it closer to some unknown reality. If the United States, with more people officially reported with AIDS than any other nation, cannot conduct a random sample of the population to estimate infection rates, what chance has Ethiopia, torn by civil war and famine?

And so the truckers carry it back and forth to Kenya and Somalia. In Nairobi virtually all the prostitutes are infected, most of them with five to ten customers per day and night, still largely unprotected despite some efforts now to encourage condom use. Recall that this was the country whose President denied the existence of HIV, saying it was all a foreign plot and a hate campaign. Five years later, even by official figures reported to the WHO, Kenya has the third highest rate in Africa, and that means in the world (excluding three or four Caribbean islands that were homosexual vacation areas in the early 80s). Nairobi, and the vacation areas of Mombasa and Malindi on the coast, are also major sources for intercontinental transmission: the tourist trade, so assiduously protected by presidential denial, is bolstered by thinly disguised sex tours from Europe, particularly Germany. As a major focus for Kenyans, especially those living in the highland areas that stretch north and west of the capital, Nairobi is a major national epicenter. People move back and forth from their rural homes to this epidemiological hub, many of them carrying the HIV with them. Rates among ordinary, low-risk pregnant women are rising, and they serve as a good surrogate measure of the infection levels in the general population.

From Nairobi, the trucking routes split, south to Tanzania and west to Uganda, and here the gradient to catastrophe rises. The western shore of Lake Victoria where both countries meet has been racked by war, from the time of Idi Amin's rampaging troops to the counter-thrusts of Tanzania's soldiers repelling them from Kagora District and the town of Bukoba back over the border into Rakai District along the lake shore. We are in one of the most highly infected areas in the world, an area of young orphaned children and old people, with many of the young and middle aged adults already dead or dying. Every year the number of orphans doubles, and there are now about 20,000 of them. In the small village of Kishenyi, one grandfather was trying to take care of 15 orphaned grandchildren, but the small farm had long ago reached its limit. Children are tragically vulnerable, for unlike war, when one parent may survive, the HIV usually takes both man and wife leaving the children without support. In areas raped and pillaged by undisciplined and rampaging troops, over 30 percent of whom were infected at the time, the human presence itself is threatened. Whole areas are evacuated by devastated people, who often carry the virus with them to new locations, and the effects can even be seen on satellite images as villages and towns are abandoned, allowing the bush to grow back over the roads and tracks visible at these resolutions. In the town of Kasensero, over half the people are dead, and nearly all the rest infected. Here, and elsewhere in Tanzania, the officially reported AIDS cases bear no resemblance to the truth, not because of official lying, but simply because there is no capacity to diagnose and register deaths. At the beginning of 1991, Tanzania had a reported AIDS rate of nearly 0.2 per *thousand*; five years before, in 1986, low-risk blood donors in Dar es Salaam had an infection rate of 7.3 per *hundred* (percent). The latter is over 400 times the former, but even to compute a national rate assumes you have a reasonable estimate of the total population of the country. Tanzania has not had a census since 1967, while the last reasonably reliable one for Uganda was also that same year.

Do you see now why it is *scientifically* proper to throw away the supposed precision of reported statistics, hold onto our commonsense, and extrapolate from a few known "spot heights" to limn the topography of the pandemic? Science is not spurious precision: science is doing your damnedest with what you know and what you can reasonably infer. And what we know are a few, reasonably accurate, spatio-temporal "spot heights," rates of infection among particular groups of people, at particular times, at particular places. But we also know, as geographers, a great deal about how things spread in space and time over a map strongly structured by human beings. That metaphor of a "spot height" is not an idle one: our map (figure 7.1) *is* the crude, estimated topography of the pandemic, the high peaks, the connecting ridges and the low valleys of infection extrapolated from what we know, a configuration thought through from a few accurately surveyed points, taking into account when the infection rates were measured, and what has almost certainly happened since in the light of distressingly general trends. It is a devastating picture, and it will get worse over the next decade: the black on the map around the spot heights will enlarge to the medium gray tone; this will spread to the light gray, and this in turn will cover and enclose the areas still relatively free. Reasonable estimates indicate that Tanzania already has 800,000 infected people, and Uganda 1.3 million. AIDS cases are doubling every month, or so we think. Kampala is the focal point, but Amin's troops did their foul work well all over the country, particularly in areas whose people did not traditionally supply many men for the military. In Rakai District the food supply is endangered simply through a lack of labor, and this is a growing theme all over this region where Uganda, Tanzania, Rwanda and Burundi meet. Even in Tanzania's capital, Dar es Salaam, where we might expect educational efforts to be the most concentrated, there is often a total disregard for the dangers involved. For many young men the acronym AIDS stands for "Acha Iniue Dogedego Siachi," meaning in Swahili "Let it kill me because I will never abandon the young ladies."

One particularly insidious aspect of the pandemic, reported frequently from Kenya and Uganda, is that as educational efforts in the towns meet with some partial success, and middle aged men begin to realize the real and quite personal danger of intercourse with their usual "girlfriends," they start to prey on younger girls from 10 years old to their early teens, knowing that these children will almost certainly be free of infection. Posters in girls' schools warn of "sugar daddies" who will tempt with sums of money that may imply some form of meaningful survival. One 14-year-old supplying sex in a Kampala bar was using the money to pay her school fees. As a result of such child sex, Uganda raised its legal age of consent from 14 to 18 years old, but there is no record of anyone actually being prosecuted under the new law. Since the sugar daddies are likely to be untested, but infected from previous contacts, the virus works its way down into the next generation.

So infected has the late teen–young adult cohort of the population become in Rwanda, particularly in the capital Kigali, that the blood supply itself became a major source of HIV infection. Secondary school children were asked to come forward to donate blood on the grounds that an estimated 18–30 percent of all the adults were seropositive. So severe was the infection rate that many teenagers attending schools in Kigali asked to be returned to their homes in the rural areas where the rate of infection in the general population was "only" 4 to 5 percent. Condom use is virtually unknown, and the reason is particularly and poignantly cultural. Much of Rwandan society is structured by a cultural framework, a way of looking at the world, that sees everything in terms of fluid flows. Traditionally the king's body was seen as a flow of fluids, nurtured by enormous quantities of milk and beer, and purged with a drastic diuretic every morning, while rainfall was again seen as a fluid flow from *Imaana*, the Creator. Drought was appropriately seen as a blockage of fluids that needed to flow but could not. All celebrations and rites of passage are similarly marked by generous offerings of fluid gifts, much as a mother offers a bountiful breast to her child. In this world, sexual intercourse is seen as a mutual and pleasurable exchange of fluids, in which the man brings the woman close to orgasm by stimulating her clitoris with his penis before penetration, producing copious vaginal secretions. A condom blocks this reciprocal flow and turns the man's semen back on itself. Most Rwandan women fear condoms for their blocking power, and are afraid that they will become lodged in the vagina. For many it is impossible to conceive of making love in such a manner that contradicts their most fundamental way of thinking about the world.

To the west of Rwanda and the Rift Valley lies Zaïre, the third largest country of the continent, with major truck routes to the south, and the one we should have the clearest picture about since it has received tens of millions of WHO and CDC research dollars during the first decade. Unfortunately most of these were frittered away in classical, repetitive, unimaginative, and geographically illiterate epidemiological studies in the capital Kinshasha and a few other towns. As a result we know a few things in excruciating detail about a small urban elite, assiduously and repetitively reported in medical journals, but little about rates of infection over vast stretches of the country. This is a typical problem, and one we shall look at more closely (chapter 11). Attempts have been made to educate about safe sex and to distribute condoms, but many Zaïreans joke that SIDA, the French acronym for AIDS, really stands for *Syndrome Imaginaire pour Decourager les Amoureux* meaning the imaginary syndrome for discouraging lovers. Nevertheless, infection rates in Kinshasha seem to have leveled out at 8 percent, as opposed to five times that in some other cities, particularly those of the copper belt of the south, where expatriate mining experts and managers have left in droves for fear of any medical problem requiring blood transfusions. It is worth noting in passing that all American embassy personnel in Africa today are flown out immediately for any intrusive medical work, including dental injections. The result is that copper production in the Katanga region fell precipitously, and combined with earthquake effects in Chile, and copper mine strikes in Peru, the world price of copper rose nearly three times between 1987 and 1989.

The truck routes on the western side of the Rift Valley lead straight through the Katanga copper belt to the equally rich area in northern Zambia. So severe is the growing rate of infection among copper mine workers and smelters that one doctor raised the question of how long the industry could keep going in the face of the pandemic. Zambia's copper belt workers are the industrial elite, well-trained, stable, often married with a family, and with above average healthcare. Education levels are also comparatively high, and support is provided for those who want to go on to further education after high school. Yet half of the 5,000 cases of AIDS reported to the WHO came from the copper belt, with most of the others at the capital of Lusaka and in the smaller truck stops strung like beads along the main north–south route bisecting the country. It is this route that leads in turn to Zimbabwe, a country with some of the highest rates on the continent, with over 10,000 cases estimated, and infection rates of 20–40 percent in the main cities such as the capital Harare. Condoms are distributed by major firms in pay packets, but this is a country in which a woman's security still lies in childbearing, and the "roora" or "lobola" – the bride price – must be paid back if no children come from the marriage. Condoms have little chance.

Towards the southern end of this long trans-African highway, bifurcating and rejoining itself like a great braided stream, lies the country with the highest reported rate of HIV infection in the world. Malawi probably has general rates of well over 30 percent in the adult populations in Blantyre and other towns, estimates compiled from blood donors, women coming to ante-natal clinics, and people undergoing testing when applying for life insurance. But the high infection rates are not confined to the towns: rural Malawi has long been a major source of migrant workers, particularly to the mines of South Africa. By 1988, HIV infection was 10 percent, and if we conservatively estimate a doubling every eighteen months (other estimates say every eight months), it means that nearly half the mine workers returning to the rural areas of the country are infecting the wives and women who stayed behind. What the year 2000 will bring no one knows, and it is even difficult to speculate. Malawi is essentially agricultural, and as in most Third World countries it is labor-intensive. How do you feed yourself with no one to work the fields, even if you make radical shifts in the crops you raise, away from labor-intensive maize to more flexible and less-demanding bananas?

And so, like two arrows coming together, the truck routes from Zimbabwe and Malawi converge on war-torn Mozambique, a country in chaos where faction has been fighting faction with modern automatic weapons, continuing the slaughter that formed the death throes of the oldest colonial regime, the Portuguese, the first to come to Africa, and the last to go. It is virtually impossible to establish even a few reliable spot heights on our map: for low-risk people in the capital, Maputo, rates may be as high as 12 percent, with similar rates for other major towns like Inhambane, Beira and Mocambique City, and presumably all the truck stops along the way. In displaced persons, HIV rates are probably 7–10 percent, and 9 percent of all Mozambique's people are refugees, with five million more homeless in transit camps. In transit to where? In the uncontrolled military of all factions the rates probably match those of other central and east African countries, perhaps 50 percent or more. The tragedy is doubled by the strong and increasing presence of HIV-2 from West Africa. As we shall see, Guinea-Bissau is one of the most heavily infected countries in West Africa, and it would appear that the old colonial connections from the Portuguese are still intact, not the least through trade and language ties. The HIV-2 hopped around the coast of Africa to the low-risk urban populations of Angola, and now to Mozambique.

And here I must inject a highly personal note. Mozambique was liberated by FRELIMO, led for many years by Eduardo Mondlane, until he was murdered in Dar es Salaam by Salazar's henchmen sending him a parcel that exploded. We were students together, taught later at the same university, and met for the last time in Dar es Salaam shortly before he was killed. He, and those around him, had such high hopes for the future in a free and independent Mozambique. He was her first and only PhD after 400 years of the *assimilado* policy, a policy of granting Portuguese citizenship to any African who could "qualify." The road to independence and its aftermath has been a complex one, but it demonstrates the fragility of newly emerging states when the few responsible leaders are wiped out – by parcels, guns or AIDS.

Only South Africa lies farther south on this great alignment. The pandemic has already begun: 3 percent of the black population of Johannesburg is probably infected, and even the WHO estimated nearly half a million cases across the country, mainly in the north and east. Migrant workers from highly infected areas of Zimbabwe, Zambia and Malawi are required to take an antibody test, but many do not. With so much "leakage" over the borders, the HIV is doubling every six to eight months, and HIV infection may reach 18 percent among the Kwazulu around Durban by 1992. The truck route from Malawi is now known as the Highway of Death: 92 percent of the truck drivers visiting Durban were infected, sleeping with prostitutes there or at stops along the way. Condom advertisements are not allowed on billboards, in newspapers, or on TV, and in any case they are regarded with grave suspicion because many Africans consider them as ways of controlling their population. Control of the epidemic is not helped by 14 different government health departments to serve the human family fragmented culturally and geographically. The newly emerging political ramifications do not help either: the African National Congress (ANC) resisted the testing of its Umkhonto guerrillas from Angola, so another dangerous source of infection has been injected into an already "on the brink" situation. Poverty and the breakdown of order in the large slums is equally dangerous: "jack rolling" by urban youths – seemingly random rape and robbery by teenage gangs – only adds more sexual connection to the South African backcloth.

Despite a leapfrogging of HIV-2 around the coast from one former Portuguese colony to another, the HIV-1 still dominates the great north–south alignment of East and Central Africa. West Africa reverses the presence of these two viruses, being dominated by the HIV-2, although the HIV-1 is gaining ground steadily. It is here, in West Africa, that I want to exemplify a second, but very differently structured alignment, an alignment focusing upon a single city, Abidjan, the capital of the Ivory Coast (figure 7.2).

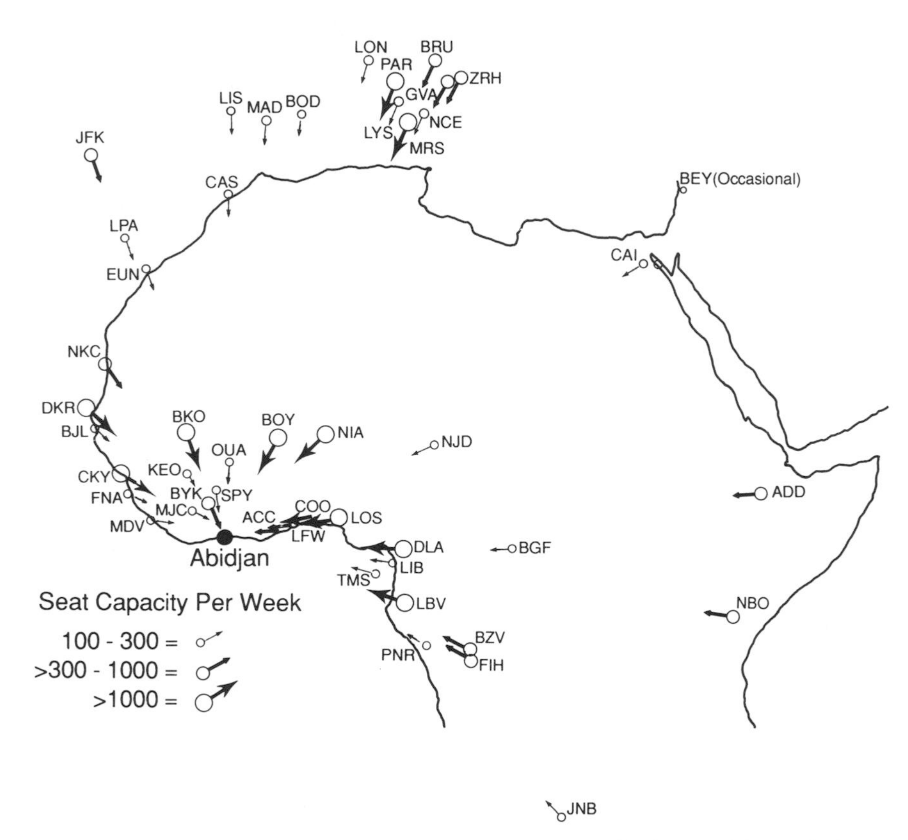

Figure 7.2 Abidjan, the capital of the Ivory Coast, as the focus of direct flights in a piece of the international air network. The city is highly connected in the international backcloth, allowing both varieties of HIV to be transmitted quickly.

Do not misunderstand me: the truck routes with their drivers and migrant passengers are still the same force in the geographic spread of the pandemic, but superimposed on these land routes is the modern air network of Africa. It is used, virtually by definition, by the modern elite, a group rich from business or government sponsorship, generally well-educated; in brief, the engineers, doctors, military officers, and, particularly and especially, the politicians.

Let us recall that the AIDS pandemic in Africa is still overwhelmingly an urban phenomenon, with the highly infected regional epicenters driving the seepage of the virus into the surrounding umlands. Let us also recall that HIV infection tends to be particularly high in the wealthy, educated, influential and politically powerful, which in most African contexts means the young and rising politician relating in casual sexuality to an entourage of willing helpers, the doctor of medicine enjoying nurses in the same hospital, and the businessman relaxing, with any one of a number of girlfriends receiving tokens of his esteem. Under these conditions the sexual relation structures the elites of Africa very tightly, and it is hardly surprising that such anticipatory phrases as "the hollowing out of Africa" are being used. Already over half, and in some cases virtually all the officers of the armed forces of east and central African countries are infected, and when highly trained and disciplined officers disappear the corporals come to power. From Germany to Uganda we know what happens when they promote themselves.

It is precisely the power elite that has access to the air network of Africa, making business trips, exchanging army and airforce officers to coordinate defense exercises, requiring politicians to attend conferences accompanied by entourages of civil servants, expert advisers and private secretaries, sending experts of all sorts – doctors of medicine, agricultural engineers, academic scholars, television broadcasters – to meetings in order to "coordinate." Often they are the sort of people that Arthur Koestler called "The Call Girls" in a book of that title, those at the top of the international circuits who flit from one conference to another coordinating all the way, usually at the expense of the taxpayers – the cocoa farmer of Ghana, the coffee grower of Tanzania, the groundnut farmer of Senegal – anyone whose income can be counted as it is shipped for export. In Africa it is always the productive poor who pay for the coordinating activities of the elite. High rates of infection, and access to rapid transportation linking the upper levels of the urban hierarchy of Africa, have proved to be a deadly combination. If ever there was a mechanism for hierarchical diffusion of a virus this is it. We shall focus on Abidjan in the Ivory Coast, but lifting this particular piece of the air network out of the dense criss-cross of flights covering the continent could be repeated for scores of other prominent cities.

Abidjan is an attractive site for coordinating and information exchanging conferences of all sorts. Loyal to France upon independence, and without a military coup in the intervening years, the Ivory Coast has enjoyed a benign paternalism and a resulting stability that is the secret envy of many other francophone nations in Africa which are experiencing the effects of self-crowned emperors, petulant marxists, sundry one-party dictators, and a succession of army generals, usually self-promoted. As a result its shops are full of desirable imported goods, even if the national debt does rise each year, and stocks of champagne, French perfumes, the elegant dresses of fashionable *couturières*, and other useful tokens of esteem are plentiful. It draws the elite of francophone Africa like moths to a candle flame. In any one week there are *direct* flights from 46 of the world's cities, 33 from the African continent, and 24 of these from francophone Africa. Of the 13 direct weekly flights from Europe and North America, six are from French-speaking cities. Even at this geographic scale the language that makes us human structures the space in which we have our being. The result is that Abidjan sits like a great spider in the middle of a web of direct connections, many of which are focal points in turn for other links funneling the francophone elite towards it. It is also known as the "the sexual crossroads of Africa."

With such a highly connected node in the urban backcloth of the continent, it should come as no surprise that the HIV-1 of East and Central Africa arrived in the early years of the pandemic. After a busy day of coordinating, the political, economic and other expert male elites need rest and relaxation, and other sorts of call girls and *femmes libres* are always available to help. By 1987, 24 percent of Abidjan's prostitutes who were actually tested at a clinic for other sexually transmitted diseases were seropositive for HIV-1, exceeding the rate for HIV-2, the variety indigenous to West Africa. The latter was already in the blood donor and hospital population at the rate of 1 percent in 1966, as we know from 207 samples of stored blood sera. Despite the HIV-2's headstart, the rate for HIV-1 had overtaken it in roughly six years. By 1989, half the prostitutes were infected, a majority by both HIV varieties, but by that time the viruses had spread both to low-risk women coming to hospitals for pre-natal care (12–14 percent), and down the urban hierarchy to other people "up country." Towns like Bouake and Korhogo, both with direct internal flights to Abidjan, reported rising rates in hospital patients, in some cases from zero to 18 percent in two years. In Abidjan's teaching hospital, associated with the university's medical school, 60 percent of the patients were seropositive to either HIV-1 or HIV-2, 30 percent of them to both. In the city hospital, HIV infected patients came from over 20 countries, and in one of them, Niger, all 25 men diagnosed with AIDS in Niamey had lived for many months, even years, in another West African country, usually the Ivory Coast. By that time 6 percent of the prostitutes tested in Niamey were infected.

Unfortunately, and despite the catastrophic rise in infection in the mid- and late-eighties, the mass media still displayed a reticence that had deadly implications. Unlike Uganda, where the courage of a Philly Lutaya showed the way to mass public education with popular songs, the newspapers and television of the Ivory Coast covering the funeral of the popular television personality Roger-Fulgence Kassy, an event attended by thousands of young people, simply noted he had died of a "long illness." Not only had he died of AIDS, but he obviously infected, or was infected by, a number of friends around him who also died. It was a highly public occasion which could have been used for forthright education instead of being stifled in silence.

Wherever we follow a connecting link out from Abidjan to another West African country we find the same pathetic story – the more virulent HIV-1 entering and gradually overtaking the slower and possibly less virulent HIV-2. Again, there is no implication here that Abidjan is a source area; it is simply that we have chosen to focus on this hub in a highly interconnected system of two-way flows. In rural areas of Benin, for example, the rates were already 7 percent for HIV-1 by 1989, roughly the same as the rate in prostitutes in Cotonou two years before, a time when their sisters in Bamako, Mali, were 25 percent infected with both varieties. The rapid spread of HIV-1 in West Africa is astounding, and in a sense our "spot height" measures of the HIV traffic give us some idea of the enormous connective power of the national, urban, urban-to-rural, and, ultimately, the person-to-person backcloths that form the geographic and human structures upon which the HIV lives and moves.

No matter what the scale at which we observe and think about the pandemic in Africa, there is little more that one can say, either as a geographer, or simply as a powerless human being. In the absence of a vaccine or a cure, how is it possible to stem the tide of HIV that is washing over the continent in a huge tidal wave? We pin our hopes on education, but those who look at the situation from the outside, with little or no experience of the conditions or the human reality of Africa's rural areas, towns and cities, cannot possibly have much sense of the overwhelming magnitude of such a task. Literacy rates in many areas are still low, and printed materials have little or no impact, even if there are reasonably effective means of distributing them – and often there are not. A typical scene would be a Land Rover from the city stopping in the only street of a rural village or small town, perhaps the focus for individual family dwellings scattered widely over the landscape. Two young men, perhaps occasionally a woman, might address the small crowd that bothers to be curious and stays to be harangued for ten minutes. A few pamphlets and posters are handed out, even if few can read them, perhaps accompanied by a few packets of condoms, probably the first that these rural people have ever seen, sources of amusement rather than savers of lives. The "big men" from the city climb back into their vehicle, start it up, wave goodbye, and are off in a cloud of red dust to perform the educational charade in the next small center down the road. The people drift away, the children blow up the condoms shrieking with laughter, and nothing changes. As education it is useless, a gesture of despair, a fist shaken at the wind. Here and there, in the cities where the HIV is already deeply entrenched in blood, brain and human cells, the effort may become a little more than just a gesture, but it touches only a few percent of those at risk.

And behind these barely effectual gestures lies poverty and ignorance that translates either into bravado or resignation, both of which may point to an acceptance of death that is difficult for a Western child of the Enlightenment to accept. We are doers, shakers and movers; we view disease as something to cure and overcome, and death as an end we pretend not to think about even as we try to prolong its approach. We wear safety belts: the people of rural Africa climbing into countless "mammywagons" have none. When you are poor and surrounded by many sources of death, when your chances of reaching the age of five are one in two, and after that the odds of reaching 40 are one in three, you tend to leave the worry about the HIV to someone else. People, either from your own government or foreigners, can throw money at you, and you will gladly, even bemusedly, take it – but not seriously, not believing it will really do anything. You are certainly not going to put aside the exquisite and natural pleasures of making love, with an average of 20 partners (a value computed by one of those Western children of the Enlightenment) before you convert to AIDS.

It is difficult for this Westerner to convey what many would be tempted to call fatalism – the sense of acceptance, resignation, bravado, disbelief, tolerance, even amusement with which a discussion about the pandemic is often greeted in Africa. And why not, Western child? You see we don't have disposable needles, or sterilizers, or condoms, or AZT, or Western blot tests, or virus-free blood, or pentamidine sprays, or hospital beds, or doctors, or nurses. We don't even have ribbons for the typewriters, to type the forms we don't have anyway, for reporting deaths from AIDS that were never diagnosed, to those who keep track of such irrelevant numbers in Geneva.

Let me end with a personal note. In December, 1988, I lectured to the AIDS Surveillance Group of the WHO in Geneva about ways of modeling and predicting the geographic spread of AIDS. While I was there I examined a number of proposals submitted by African countries to obtain funds to combat the pandemic. The invitation to submit had been extended by the WHO, and there were literally scores of these proposals from all over the Third World, all neatly indexed and alphabetically arranged in a special and quite spacious storage room. When I opened Sierra Leone's proposal I knew I was back in Africa: the thin sheets smelling of laterite dust, the unaligned type from an old typewriter heavily indented to make three or four increasingly faint and illegible carbon copies. The usual preamble, Edwardian in tone, "I have the honour to submit . . . most obedient servant, etc.," and then the proposed budget, a *national* budget, to fight one of the worst pandemics in human history. And there it was, as an individual item in the equipment needed to overcome a virus that even then was in 5 percent of the blood donors of Freetown. Three bottles of Tipex . . . estimated cost, US $2.10. A white correction fluid for erasing typing mistakes, next to two reams of paper, and a box of typewriter ribbons. If, of course, you can spare the money.

8

Thailand: how to optimize an epidemic

Thai men need to cultivate good hobbies so they have positive activities for their time and energy.
Dr Saisuree Chutikul, sole woman member of Thailand's cabinet, 1991

I am afraid that for behavior to change, we need many more Thais to die.
Vicharu Vithayasai, Chiangmai Hospital, Northern Thailand

What are they going to do when they get to the last woman?
Chantawipa Apisook, of Thailand's EMPOWER

If Africa is already a continent in catastrophe, Thailand is a nation rapidly approaching the same chronic state. So quickly has the HIV spread that the country may soon overtake Malawi and Uganda as the most infected country on earth. But first a caution: in moving from Africa at the continental scale to a single nation in Asia we should be aware that we have radically changed our geographic scale. We are going to be looking at Thailand through a geographic "lens" that magnifies details over 50 times compared to the one we previously used in Africa. We are going to examine in considerably more detail a country roughly the size of Kenya. It is important to realize this because the estimated map of HIV infection in Thailand (figure 8.1) superficially resembles the continental map we looked at before: once again the HIV is an urban phenomenon with seepage along the major highways and out into the rural areas. Again we only have "spot heights" of HIV infection, often for selected and high-risk groups in the population, but conservative and essentially commonsense extrapolations from the time of reporting, and over space from specific places and areas, gives us a crude approximation that is probably a reasonable picture at the end of 1991 when the map was compiled. If there is some slight exaggeration in some areas, they will be underestimates by the time this book is published. The virus is moving that fast on a social backcloth that is probably more tightly structured by sexual relationships than even many cultures of Africa.

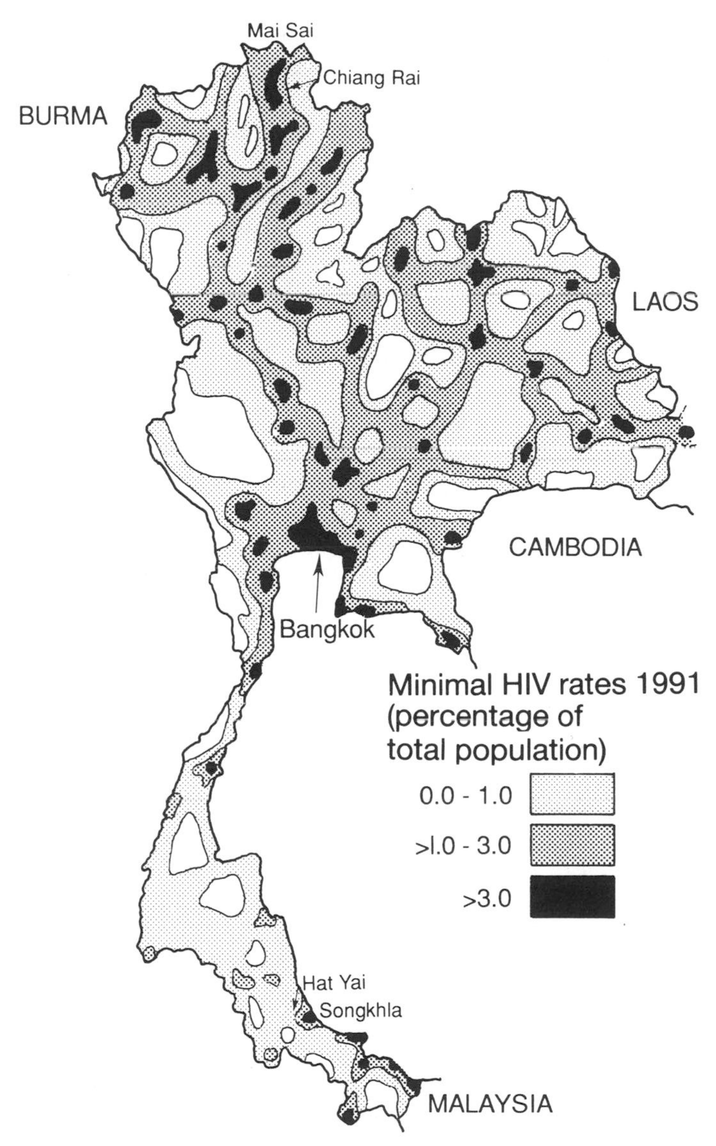

Figure 8.1 The generalized map of HIV infection in Thailand around 1991, compiled from many sources. As in the case of Africa (figure 7.1 p. 74), the estimates are conservative.

Like many maps displaying a piece of the human condition at this scale, there has to be a fairly high degree of generalization. The urban areas (black) stand out as peaks where over 3 percent of the people are infected, but in some of them 90 percent or more of commercial sex workers (CSWs) have the virus, and we are in a culture where there are 500,000 to 800,000 of them, or roughly 10 percent of all women between the ages of 15 to 24. As we shall see, there are particularly high rates in and around Bangkok and some of the "recreation towns" immediately to the south-east of the capital, in the north close to Burma, along the north-eastern border with Laos, and in the far southern tip of the peninsula near the Malaysian border. The dark gray areas of moderate infection closely follow the main roads, those of the north following large south-flowing rivers, while in the north-east they cut across a series of large tributaries flowing east and north to the Mekong forming the border with Laos. Both road systems focus or "drain" the country towards the capital. If we could magnify a stretch of one of these major arteries 50 times again, we would see a series of small towns and villages likes beads on a string, many of them truck stops with high rates of infection. Virtually every one of them has a bar-brothel or two patronized by local workers, farmers and truckers, and there are 200,000 of the latter in the country, most of whom engage regularly in unprotected sex with CSWs who are anywhere from 30 to 90 percent seropositive. Once again, major lines of transportation are built to serve people, and it is people who carry the HIV. Distance from a main road has only a small buffering effect for the rural areas, since many small agricultural towns served by small feeder roads have their own brothel-bar catering to the local farmers and workers. There are few areas untouched by the HIV (white areas on the map), and they are mostly regions of difficult terrain, no motorable roads, and few people.

The map of estimated HIV infection is literally a superficial geographic expression, a single measure amplifying with distressing speed, lying on the surface and exposing to view a glimpse of much deeper cultural complexity. To understand how a virus can spread so quickly and with such terrible effects we shall have to dig beneath the hills and valleys of our contoured "infection surface" starting a couple of hundred years ago. Traditionally Thai village life was marked by relative equality between men and women, and the line between men's and women's work was flexible. Both engaged in agricultural labor, trade was often in the hands of women, and inheritance was an equal division between sons and daughters, with the expectation that the youngest daughter would receive the family home. In marked contrast, urban and court life was almost totally male dominated, supported by Brahminic notions of a king divinely ordained that pushed aside any gentler Buddhist ideas that happened to be inconvenient. The little that most people know about Thailand today is probably strongly colored by the American musical and film *The King and I*. Subtracting the obvious and entertaining theatricality, and recalling its basis in a diary of an English governess of Victorian times, the view of male dominance in the court and upper reaches of Thai society were probably close to the truth. To this day, *The King and I* is banned in Thailand on the grounds that it shows disrespect for the Thai nobility.

During the late nineteenth and early twentieth centuries, male domination increased significantly. Any degree of Westernization meant imports, and imports required exports in return to pay for them. Rice production became highly commercialized, often by Chinese and other immigrant groups whose own male-dominated activities required the cooperation of government bureaucrats and government officials – at a price, of course. This elite was joined in the 1950s and 60s by a rapidly rising commercial class whose male members tended to believe that power and money went hand-in-hand with wives, consorts, concubines and other women-for-pleasure. Rising commercialism in the nineteenth century also meant increasing internal migration, particularly of young males to urban areas, so that prostitution expanded even as multiple sexual liaisons were legitimized by the behavior of the rich and powerful. We shall see many ramifications of all this as we try to tease out the numerous strands forming the sexual backcloth on which the HIV lives and moves. Migration is still extremely strong as people leave the poorer rural areas for the cities, but it means different things for men and women, and it is probably grossly underestimated because the Thai census is based mainly on house registrations. In the 1980s, for example, 90 percent of the recorded net migration was to the Bangkok metropolitan area.

As far as observable conversions to AIDS are concerned, the HIV was relatively late in arriving in Asia and Thailand. The apparent delay was such that in the early 1980s some Asians were even saying they had a natural immunity to it. In Thailand the first reported case and death only came in 1984, a homosexual man who had spent many years in the United States. This was reassuring to many: it was going to be a disease of *farangs* (foreigners), homosexual men and intravenous drug users after all. The latter were already showing signs of infection, not the least because there was a potential pool of 200,000 heroin users, 77 percent of whom shared needles or got direct fixes from numerous dealers in every major city, particularly Bangkok. Heroin users and drug dealers in jail at the time of the king's sixtieth birthday in 1987 were all released in an act of clemency, but most of those who had been HIV free on entry to the jails emerged infected. Most of them promptly spread through the city, got their fixes at the usual places, and so passed on the virus to numerous others. The following year the infection rate in IV drug users reached 15 percent and then soared to 43 percent in the next six months. It now appears to have stabilized at 35 percent with some education and the provision of free bleach to sterilize needles.

The year of the king's birthday, 1987, was also declared "Visit Thailand Year", a culmination of an intensive campaign of publicity enticing tourists to the Land of Smiles. So successful was the campaign that it was extended for six months to make it the longest year in history. Throughout the seventies and eighties, official tourist publicity – pamphlets, posters, spot television advertisements, etc. – painted a land of beautiful young women with coy "come hither" smiles, and the sexual message became so overt and obvious that the government told the Tourist Board to tone it down and emphasize tropical beaches *sans femmes* and temples instead. But after "Visit Thailand Year" there was no further need for sexual publicity: Thailand's commercial sex industry was famous or notorious worldwide, and by 1990 Thailand was attracting 5.3 million tourists from over 30 countries, with a high ratio of single men, particularly from Malaysia, Japan and Taiwan, the main source areas. Special sex tours flew into Bangkok daily loaded with Europeans and Japanese, while Arabian sheiks chartered Gulf Airways planes for themselves and their large male entourages. A return flight was often dubbed "the gonorrhea express," since 70 percent of the CSWs in some of the more popular brothels suffered from venereal diseases. In the days when a shot of penicillin could cure, and before super strains resistant to antibiotics appeared, many clients considered it worth the price. Prices varied, of course, but one German company provided clients with a catalog of available women, two weeks in a tropical paradise, and a round trip ticket for $3500, about the same price charged by similar travel agencies in the United States, including those arranged by one enterprising male agent who called himself Captain Sticky.

In an international economic system, where the fine hand of the market rules, entrepreneurial opportunities quickly become apparent. By the late seventies, one prominent brothel owner in Hamburg linked up with Charlie's Assistance Office in Bangkok, and was able to deliver young Thai women to German, Dutch and Danish investors in commercial sex for $1000 each. A slave market (there is really no other word for it) was opened in Frankfurt where a young Thai woman could be rented for $2500 per month plus a $1500 returnable deposit. Placed in prostitution, she would earn about $275 a day, to provide a return on investment of approximately 300 percent per quarter, considerably better than the stock market or certificates of deposit in German banks. Charlie's in Bangkok operated with the full cooperation of the Thai government to provide papers, visas, etc., while at the German end extended rental agreements could result in outright sales of women for $10,000 each. When German police started to investigate, the women were rotated to other brothels and cities, homosexual men were paid $5000 to go to Thailand to marry Thai women and then dump them on arrival, while the flow switched to Amsterdam and Zürich as gateways to Germany.

By "Visit Thailand Year" evidence of the coming HIV epidemic was already quite apparent: HIV was in the CSWs and IV drug populations of Bangkok and other major tourist centers, but the government decided not to launch an anti-AIDS campaign on the grounds that it did not want to "incite panic." Two years later, 0.5 percent of all pregnant women (non-CSWs) coming for pre-natal care were seropositive, with some northern provinces reporting over 3 percent. Already the chain from intravenous drug users to CSWs to male clients to wives and children had transmitted the HIV over the highly structured backcloth of Thai society. By this time nearly 10,000 HIV carriers had been officially identified, but responsible doctors estimated that this represented only about 10 percent of the total. A number of women's groups declared "Thailand is No Sex Land!", and they attempted to meet sex tours at the airport with signs of protest. They were met by a solid wall of police who forced them out of a back entrance, the same treatment they received at Pattaya when they tried to protest peacefully with signs as American sailors came ashore for local leave.

Four years after "Visit Thailand", in August 1991, the Prime Minister declared that he might have to rethink the national AIDS policy, noting however that "we don't want to create panic." At that time, doctors estimated more than 300,000 people were infected, and AIDS rose to sixth place on the priority list of government social policy. But as one doctor noted "What's the use of trying to be first in the fishing industry when fishing crews are suffering from AIDS?" In October 1991, the Ministry of Health declared it had identified 171 AIDS cases, while the previous July it had discovered 364, of which 119 had been officially reported to the WHO. In the same year, the opinion page of a Bangkok newspaper noted that AIDS was "most commonly transmitted through anal intercourse and intravenous drug use," at a time when over 40 times more heterosexual than homosexual cases had been officially identified – for whatever such official figures were worth, since under-reporting by a factor of 20 was going on in Bangkok alone. One cabinet member closely associated with family planning declared plaintively in an anti-sex tour speech "How could the *farangs* come up with certain men who exploit uneducated women?", and asked all brothel owners in Thailand to close for one day, 1 December 1991, as a token of respect for World AIDS Day. This in a country where for the two previous decades sex tourism had been encouraged, where prostitution is illegal, where one study estimated 450,000 Thai men visit a brothel every day, and where the Chief of Police has declared repeatedly "There are no brothels in Thailand." The swelling hills and peaks of Thailand's HIV surface would seem to be propped up by hypocrisy and contradiction as much as anything else. What is going on here?

It will be tempting for those who choose to see the world exclusively through neomarxist spectacles to blame a wicked global capitalist system for all the human ills of Thailand. Certainly there is much that can be laid at the door of a materialistic exploitation that reifies people into commodities. But for those who wish to see with clearer eyes the issue is not so simple, and "blame" is a word inappropriate in any epidemiological context. As we have seen repeatedly, in our closely-knit modern world a deadly and slow virus arrives unannounced and unknown in one way or another. It is the structure, the connectivity of the backcloth it finds upon its arrival that determines its ability to exist and be transmitted. It is difficult to think of any society in the world today more tightly structured by commercial sexual relationships than Thailand, and to understand the geographical dynamics of this virus we must get at least a sense of what is involved.

Estimates of the number of CSWs in Thailand, a nation of about 57 million people, range from 200,000 (the official government estimate) to 800,000. Most non-governmental, medical and women's groups put the number around 500,000. With numbers like these I am not sure that 100,000 one way or another is particularly pertinent. Most people aware of the sex industry in Thailand appear to believe it is mainly in Bangkok and a few other major tourist towns and resorts. What many do not realize is that the brothel, however disguised as a sophisticated city bar, a sports arena for Thai boxing, a "short time" hotel, or a roadside truckstop, is an indigenous institution that is found all over the country from the largest and most cosmopolitan city to the smallest agricultural village. The Red Cross estimates conservatively that Bangkok has 40,000 CSWs; Mai Sai, a small border town and crossing point into Burma, has 62 brothels and 30 broker agencies; and a small village near Chiang Rai boasted seven brothels, each with 4–12 women, serving an average of 15 clients per day (the record, told with some pride, was 76), mostly young farmers and local workers, including schoolboys still in their uniforms. By the age of 15–6, 90 percent of Thai boys have had their first sex with a CSW, a practice often introduced by fathers on the birthday after the first faint signs and sprouting hairs of male puberty.

By 1991, 1.5 percent of male secondary education students were HIV infected.

Since the brothel-bar is an expected and accepted institution of the countryside, distance from a major town that might serve as a delaying buffer from a regional epicenter has little protective effect. Whatever collective noun we may adopt for CSWs, mobile groups of these tour the rural areas to serve market fairs and market days at smaller, more remote towns and villages. As material prosperity spreads into a rural area, so new brothels are opened to supply a demand formerly unmet. For most Thai males, sex is simply regarded as a casual pleasure: one teenage male, who became HIV infected, offered rides home daily on the back of his new motorbike to young women working in a rural factory nearby. His passengers returned the favor by inviting him in for sex as a token of their gratitude.

For the foreigner, the casual and accepted pleasure in sex by ordinary young Thai men may take some getting used to. One young American on a visit to the rural town of a Thai friend was invited to come down to the local brothel one Saturday night. After a few drinks many of the young men drifted off to the partitioned cubicles at the back with the bar girls, and after a few more drinks others stood on chairs and stools watching the performance over the partition, cheering their fellows on, while the girls, legs akimbo, gaily waved back and encouraged their clients to a more vigorous public performance. For the young men, this was considered an absolutely standard and accepted evening's entertainment in this rural area, and presumably a good time was had by all. No condoms were used because foreigners never came there as a rule.

In Songkhla Province, far to the south, 8,000 CSWs are available on a regular basis, augmented by 2,000 more on holidays. Equally remote from Bangkok, a favorite army posting, soldiers sent to the 17 more rural and remote northern provinces are hardly sexually denied, for the Royal Thai Army keeps a list of 1,290 recommend brothels employing 10,241 women. The result is that 4–8 percent of the Thai military is infected there, not forgetting 5.6 percent of the police.

In a nation in which prostitution is illegal under the 1960 Anti-Prostitution Act, and bills to decriminalize it and allow CSWs to be treated as independent business women are shot down in committee before they even reach Parliament, what is happening? Many feel it is precisely the fact that prostitution is illegal that makes it so prevalent. This is too simple, but there is no question that enormous power and huge sums of money are involved, so that corruption permeates every level of the society. Many brothels are owned by prominent politicians and police officials; virtually all males involved in politics at all levels up to the Cabinet use CSWs; and police use them for free, as part of the gratuity expected for allowing them to exist. Huge bribes are given on a regular basis to senior police in all 72 provinces, and the sex industry everywhere operates under a male consortium of the army, the politicians, the government administration, and the business community. When Hat Yai, a major sex center in the south suffered a decline, the voices raised in concern revealed clearly the underlying structure of support. Hat Yai serves a large male-dominated, tourist clientele, 70 percent of it from Malaysia and Singapore. When the Malaysian government publicized the high rates of HIV infection in CSWs, and banned all its government officials from traveling to Thailand, the protests came from the general in charge of the Fourth Region (in which Hat Yai lies), the governor of Songkhla Province, the mayor of Hat Yai, the Hotel Association, the Songkhla Tourist Association, and other businessmen in the area. They blamed government ministers in Bangkok for making "irresponsible and misleading comments on the AIDS situation in the South," urged "easing in the enforcement of laws," and denied that Hat Yai had 9,000 CSWs infected with HIV. One member of the consortium even offered one million baht (about $50,000) to anyone who could prove that Hat Yai had "only 4,500 CSWs with HIV."

A sex industry on this scale, supported implicitly by the power elite in government, business, the police and the armed forces, generates an enormous support structure to supply the national demand for young women, let alone the export trade. It is illegal for a CSW to be under 18 years of age; most entering the trade are still in their early and middle teens, with 12-year-olds not uncommon, and the youngest reported in a brothel only aged nine. But there are thousands of street children in Bangkok alone, many under 10 years old, and available at night to anyone willing to pick them up and pay them. There is a vast network of brokers and agents reaching into the smallest rural villages and using the most modern communications devices and computer records. Telephone brokerage negotiations cover the Asian continent, and reach as far as Europe and North America, with the pictures of the young women available faxed to the potential buyer. While the young women come from all over Thailand, the majority are from the poorer northern and north-eastern provinces along the borders of Burma and Laos. Even these poor areas cannot keep up with the insatiable demand, and more and more young women are being sucked into the sexual cauldron from Burma, Laos, and China. By the end of 1991, even the police were reporting a large increase in CSWs from Yunnan Province, some of them only 12 or 13 and unable to speak a word of Thai.

The recruitment of young women in Mai Sai District in Chiang Rai

Province, the most northern province touching both Burma and Laos is typical. Like much of the north, the region is poor, with an agricultural economy. In the more mountainous areas, particularly over the Burmese border, life is hard and made more so by the stark contrast with the infusion of modern material life. The national electricity grid follows the roads, and with it comes the demand for electrically operated appliances – refrigerators, TV sets, etc. – as well as the strong push by government to open up this beautiful natural and scenic area to tourists. It is a mark of both poverty and the low and disparaged status of women that many of the young women from the area supplying the sex industry are sold by their parents for the sheer material gain. Many have had their lives shattered, and their livelihoods destroyed, by rampant land speculation that has driven up the price of land to 80 times its former value a decade ago. Much common land has been sold out from under the villagers by the elders, and young men who protest end up with a price on their heads. A network of sexual agents and brokers scours the region, visiting villages and farms, looking for young girls from 12 years old and up, offering an average of $500, with a range from $160 to $800, for an especially pretty girl, with a TV set or refrigerator thrown in as a bonus. In the Mai Sai District, 65 percent of the girls who finish primary school never go on to the secondary level, and of these 55 percent are sold as CSWs. One teacher of the sixth grade at a primary school said "The classroom is like a showroom for the brokers." As in Africa, with its sugar daddies seeking younger and younger girls who are HIV free, so the demand for younger children is increasing in Thailand, and reaching increasingly across the borders to Yunnan in China.

The attitude towards the brokers is equivocal: some parents regard selling their daughters as shameful, but do it anyway, lying or using euphemisms to cover the "temporary" absence of one of their teenage children. Others regard the brokers as saviors of the family, something that many of the young girls believe too. Having been brought up to be submissive and grateful to their parents, they regard their sale, euphemized or not, as a filial opportunity to pay back the support their parents have given them. Most send a major part of their earnings back home, usually between $50 and $100 each month, all that is left over after deductions have been taken by the brothel owner for paying back the initial "debt" (twice the price paid for them), food, and days lost to illness or menstruation. Many will work through the menstrual period in order not to lose income, telling the customer with a smile that his male prowess has made her bleed. The probability of HIV transmission is particularly high at this time. Most do not use condoms if the client does not want to – and most do not. Only 9 percent in 1989, and

18–30 percent in 1991, would use a condom with a CSW: after all, HIV is only a *farang* disease, and in the village only Thai men use the women. As a result, by 1991, 5 percent of 21 year old Army recruits in Chiang Rai and Phayao were infected, while in Chiang Mai, 15 percent were infected, along with 8 percent of the blood donors, and 4 percent of the women coming to pre-natal clinics. Among the CSWs the rate was 50–90 percent depending on the particular northern province, far higher than the official 12 percent national average. Across the border, a major source of CSWs from the Karen people, 75 percent of the women CSWs who had worked in Thailand were seropositive.

If HIV flows easily on a tightly connected sexual backcloth, so does money – over a billion dollars of it each year. If this seems exaggerated, let us pick it apart and see where it comes from. It starts with the male clients who generally pay the bar owner, brothel madam, or call girl agency directly according to a sliding scale that is generally higher in Bangkok and the tourist centers, falling to more locally accessible rates in the provinces and rural areas. An exclusive CSW and call girl hired by wealthy businessmen, politicians, army officers, and tourists of refined tastes in Bangkok may be priced at $100, but this is exceptional. Most CSWs working in bars or brothels will sit in a waiting area, sometimes behind a one-way mirror, waiting to be picked out. Those wearing clear plastic tags are worth $20 for half an hour, while red tags indicate $12, blue $8, and yellow $4. In the lowest paid brothels, truck stops and rural bars, the rate may be as low as $2 to $2.60. As the price goes down, the HIV tends to go up. In one brothel where the CSWs were tested, the $20 clear tag women had little HIV, mainly because they were new, exceptionally pretty and desired by those who could pay, and they could sometimes persuade the client to use a condom. Yellow tags were about 20 percent infected, while those in the lowest grade brothels and bars were 80 percent, mainly because they had been dismissed from higher class establishments.

In high grade brothels, the CSWs are meant to be tested for HIV every two or three months. In fact, and using Songkhla as an example, only 3,000 of the 8,000–10,000 CSWs of Songkhla – less than half – had ever been tested. In any case, testing is essentially meaningless: HIV infection will not even show up during the first six weeks to three months, but while a woman is asymptomatic or not infected, she will be issued a green card to show the client if he demands it. When a CSW becomes seropositive, the green card is taken away, and the woman is meant to return to her home, often a rural village in the north. On the other hand, counterfeits are easy to obtain, or a woman will get a friend to go to the testing for her. If you wanted to devise a national policy to optimize the spread of a deadly disease, it is difficult to think of a more efficient way. You bring young, often very young women from the rural areas to the HIV cauldron of Bangkok and other major sex centers, wait until they are infected, and then send them back to spread the HIV throughout the rest of the country. In fact, many women do not return immediately. They either stay at the brothel at a reduced price, or move to a lower class establishment, often promising to use condoms, a promise impossible to keep if the man does not want to use one, and 84 percent do not. Sometimes HIV positive sex workers will tell the male client that they are infected, but some still refuse to use a condom.

At these prices, what does the daily national take add up to? Suppose we estimated, quite conservatively, and balancing the $2.60 and $4 encounters in the rural areas against the $20-and-more hirings in the cities and tourists centers, an average of $6. Using Thai researchers' own and quite reasonable estimates of 450,000 brothel visits per day, we have nearly $1 billion per year. Divided by half a million CSWs, this would provide an annual income of about $2000 per year. But like the society in which it is embedded, income among CSWs is highly skewed: a very few, quite exceptional women in a high class brothel in Bangkok might earn $1250 a month, but as one detailed and thorough study noted, "a bonded girl in a grotty place with a drunken and dishonest owner would be very lucky to get 500 baht [$25]." Most women only get one-third to one-half of the fee after "costs and overheads" and profits have been taken by the owner or manager. Most of the costs end up in the hands of the police, local administrators, or senior officers of the army with the power to recommend suitable establishments for their men and to close others. This power elite extends its influence to the highest levels. One cabinet minister declared in public that the repeal of the Anti-Prostitution Bill of 1960 had been discussed in a cabinet meeting, but said "Suppose the police don't cooperate?" Ironically or tragically, many CSWs prefer the security of police protection, feeling much safer in brothels and short time hotels under the watchful but lucrative eye of the local guardians of the law. Periodically, and purely for show and publicity, often when the payments have not been up to scratch, the police will raid an establishment and "rescue" a few CSWs held in bondage. Most of those "rescued" return quickly to the same establishment rather than return home.

This, however, is only the income generated by the domestic side of the sex industry, although certainly enough to lubricate the social relations of many in the upper classes, and keep them in the style to which they have become accustomed. Thailand also supports a brisk export trade, facilitated by still more tokens of esteem to police and administrators, and supported by the very latest technology. Thai women are greatly desired in the brothels of Japan, Malaysia, Hong

Kong, Taiwan and Singapore, and, as we have seen, also by those in Europe, particularly Germany. "Little Bangkoks" are springing up in the large cities of Japan, and the trade is highly coordinated by Thai and Japanese gangs. One study noted that the international gang coordination and skill far exceeded that of the various police forces involved, many of whose senior officers gather at international meetings from time to time, pass resolutions "to coordinate," and return home to business as usual. Thai CSWs travel via all sorts of devious routes, often via Kuala Lumpur or Singapore, to enter Japan as tourists on doctored passports. Once in Japan, and under the bonded agreements and control of the Japanese gangs, they are auctioned like the slaves they have become from prices ranging from $20,000–35,000. Today, some of the older Thai CSWs who arrived in the seventies have paid off their costs and have become skilled brokers themselves, using their knowledge of both the Thai and Japanese languages to facilitate international trade, and to control new Thai women who have no one else to turn to.

Under these conditions it is very difficult for the women of Thailand to empower themselves and generate genuine reforms. It is a society ostensibly open to educated women of the middle and upper classes, but even they are embedded in a male-dominated society adept at euphemistic descriptions and skilled at dissipating any worrisome protests into bureaucratized channels where nothing eventually happens. Women are actually barred from seven administrative positions in government by either ministerial regulation or cabinet resolution. The grounds are national security. They cannot, for example, become forestry officials, district auditors or district officers, a strange mix of occupations placing the national security of Thailand at risk. The real reason, of course, is that all these are positions with the potential to generate large incomes from payments other than the prescribed annual salary, and so must be reserved for men. A woman district auditor might actually audit, while district officers have the power to control much of what happens in the area of their jurisdiction, including local commerce. The flow of appropriate tokens of esteem might dry up or not be accepted if a woman were in charge. Even when women are in charge of local committees contributing to everything from rural development to AIDS prevention, government funds and public credit are invariably given to the local administrative offices and village headmen. As for forestry, the state secrets that women might disclose to foreign spies could hardly endanger Thai security, but they could definitely prove embarrassing in an international arena of growing concern for the global environment. Satellite images show widespread devastation of tens of thousands of square kilometers of Thailand's teak and tropical hardwood forests, much of it exploited, like the forests of Brunei and Eastern Malaysia, by the insatiable appetite of the Japanese. However, the yen is strong, and cutting can always proceed by approaching senior forestry officials and other politicians appropriately. If these officials were women willing to blow a whistle or two some action to halt further devastation might actually take place. This, of course, would be totally unacceptable.

The empowerment of women is not helped by the increasing commodification of Thai society, as we have seen in the arrival of modernizing forces such as the electricity grid, and the exchange of daughters as CSWs for refrigerators. Despite a few women's groups like EMPOWER (Education Means Protection of Women Engaged in Recreation), APSW (Association for the Promotion of the Status of Women), and Friends of Women, whose members can help a few individual women at the local level, any sharp impact appears to be absorbed by the pillow of male-dominated bureaucracies at all levels. Yet more and more land, and concomitantly more and more people, are coming under the exploitative influences of the tourist trade, driven by the entwined forces of commerce and politics in Bangkok. The beautiful northern region around Chiang Mai is typical: huge hotels for *farangs* and rich Bangkok Thais are springing up, and the region already boasts seven golf courses to serve the tourists, mainly Japanese. Local employment rises, of course, but the choices for women are limited. A woman may continue in agricultural labor, handicrafts or local industry at 50 baht [$2.00] per day, or enter the resort employment pool as a golf caddie, waitress or seeder of golf courses at 90 baht [$3.60]. As a go-go dancer at the numerous bars also springing up to serve the *farangs*, she will make 267 baht [$10.68], while as a CSW she will bring in 500 baht or $20 per day. When a woman has only primary school education from a rural village, a position in a society where men consider it courteous to offer women, often local beauty queens, to visiting dignitaries, and a strong sense of obligation to the family, it is hardly surprising that the latter choice is frequently made.

What has been the official, that is to say governmental and bureaucratic response to the growing epidemic? Like most bureaucracies around the world, it has both convened and taken part in numerous national and even international meetings "to coordinate." In 1991, for example, one decade into the pandemic, hundreds of delegates from 49 countries, including Thailand, met at a cost of over $1 million in Manila under the auspices of the United Nations Economic and Social Commission for Asia and the Pacific (ESCAP), and actually, not to say boldly, "identified AIDS as an emerging health threat in the coming decade," something that apparently no one had had the perspicacity to realize until that coordinating moment. Meanwhile, back in Chiang Mai, the WHO sponsored another meeting at a cost of nearly a quarter of a million dollars, using the new conference facilities offered by the tourist hotels, to bring officials from Burma, Thailand, Laos, China and Vietnam, supplemented by 33 WHO "experts," to discuss how to stop HIV from spreading in drug addicts in the "Golden Triangle" and other opium–heroin producing areas, where an IV shot of heroin may be had by putting down 25 cents and then sticking your arm through a small window of the dispensary. It was in the same exciting month, October 1991, that thousands of delegates and their support personnel met at the annual World Bank meeting in Bangkok at a cost of millions of dollars, using a new international conference center built especially for it at a cost of $100 million. Unfortunately, the shining new center overlooked a miserably poor area of Bangkok, which might have been offensive to the finer sensibilities of the delegates. A shielding wall was constructed, but it was soon realized that the poor families could still be seen, so in the end it was easier to "relocate" them. It was at this World Bank conference that thousands of keychains with plastic snap containers containing condoms were handed out by a government minister known to all as "Mr Condom." They were a great success and most of those attending thought they were such a cute little memento of Thailand and the international meeting. Meanwhile, across the way at the luxurious Bangkok Palace Hotel, and still one decade into the pandemic, the Ministry of Health, with the cooperation and financial support of UNICEF, was holding a series of educational lectures for doctors of medicine on how "to impart knowledge on the prevention of AIDS to the public." After all, and as the Prime Minister had declared publicly two months before, "Young people should be educated." This novel idea was stated upon the advice of the newly formed National Committee on Prevention and Control of AIDS, formed "to promote, support and coordinate . . ." At the end of 1991, the USAID and Family Health International said they would support an anti-AIDS center for south-east Asia to be located in Bangkok. It will require hundreds of millions of dollars to build and operate in order "to disseminate information and coordinate."

Not that education is not still needed: a health survey found that the lowest levels of knowledge about HIV were in "villagers, sex workers, village leaders and *health personnel* [italics mine]." One owner of an imaginative brothel in Bangkok told a health educator that there was no problem of HIV because all her women were "water people." The establishment specialized in sex in large baths, and no protection was necessary because the warm water would wash the HIV away. Players in traditional traveling folk theater have provided one medium for the education message, gathering a rural crowd, putting on a traditional piece, and then acting out a new play carrying the message of AIDS and its dangers. They amuse many, but little behavioral change has been seen so far. But ignorance is not confined to the relatively uneducated. At most universities the *ruen pee-ruen nung* tradition flourishes, one which requires that older male students take the younger ones to a brothel and treat them to women. Few of the younger students dare to refuse for fear of being laughed at or ostracized, although at one university (Khon Kaen) there was a student protest to stop the tradition. In Chiang Mai, where the university has a hospital and faculty of medicine, 72 percent of the CSWs were already infected by 1989. Since many male students have regular girlfriends who are also students, the HIV is transmitted into yet another part of Thai society, one at the threshold of productive life informed by higher education.

Widespread education is also required to reduce very sharp discrimination against those with AIDS. People who are known, or are even rumored to have, AIDS are shunned, and in some cases are driven from their neighborhood or village. One security guard at a hospital was found to be infected and promptly dismissed. He and his family were forced from their house and had to leave the area. Some nurses have refused to treat AIDS patients. Reports of suicide couples regularly make the newspapers. One CSW who returned to her village, and built her parents a fine new house on her earnings, was confined to a small hut in the back when her parents learnt she was infected. A son might have been treated differently, because men are generally not considered responsible for spreading the virus. Like all sexually transmitted diseases, HIV is essentially a "women's disease," although Mr Condom himself has declared "more women will be infected than men because Thai men have several sex partners." The ignorance underlying such a statement becomes apparent by following the potential course of the HIV through ten couples (figure 8.2). If the first man becomes infected through homosexual, heterosexual, blood or IV needle transmission, he can pass it on to his wife or girlfriend and to a CSW (A) in the first "generation." On the second, she can transmit to three other men who visit her, and these transmit to their wives in turn on the third generation, including a second CSW (B). She (B) will transmit to four of her other clients on the fourth generation, who infect their partners on the fifth. The only people HIV free are the monogamous couple completely disconnected from this small part of the backcloth, and the misogynist about to enter holy orders. As before (figure 4.7), the *average* number of sex partners is equal. Even if there is some asymmetric bias to male–female transmission, say twice the probability of women becoming infected, a value estimated from a quite different

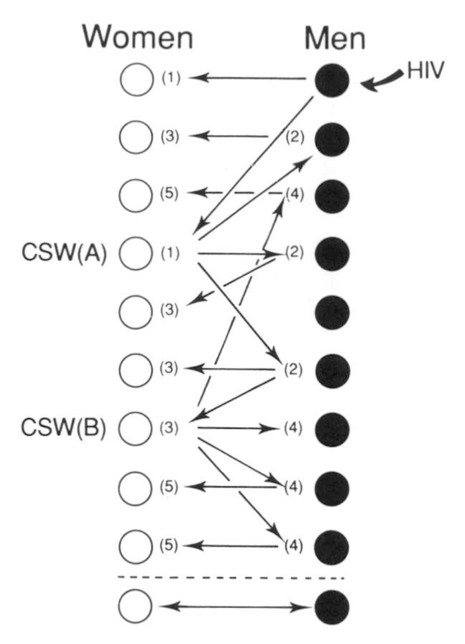

Figure 8.2 Following the transmission of HIV through ten couples to indicate that average rates of infection in men and women will equalize, despite statements by Thai officials that HIV is mainly a "women's disease." Numbers indicate successive "generations" of infection.

population in North America, the asymmetry only retards the eventual transmission. Under conditions of frequent sexual encounters, and the presence of other sexually transmitted diseases leaving open lesions, the rate of infection in men and women will be approximately equal – as the terrible experience in Africa demonstrates.

Unless disconnection occurs, there is little to prevent the leakage of the HIV from the infected to the uninfected, whether we consider a backcloth of individuals, or aggregate to whole populations at the national level. Already the HIV epidemic is spreading over a backcloth of national connections (figure 8.3), moving across Thailand's borders to Malaysia, Cambodia, Laos and Burma, and from the latter two to Yunnan Province in south-western China. By 1991, Malaysia officially identified 2,070 people HIV positive, most of them probably infected by visits to the sex centers of southern Thailand, although Thai women are also imported to serve in the bars of Malaysian towns. Authorities expelled 38 CSWs from the bars of Kuala Lumpur after finding they were infected. The strictness of a Moslem state like Malaysia seems to have little effect beyond a condemnation of condoms by religious leaders. In northern Laos, and along the border of southern China, the HIV infection rate is doubling every six months, mainly through injecting heroin. This is also a major form of transmission in Burma,

Figure 8.3 The leakage or transmission of HIV from Thailand over a backcloth of international connections to nearby countries. Backcloth structures, allowing for the transmission of HIV traffic, exist at many geographic scales.

augmented by infection rates of 75 percent in Burmese women returning from brothels in Thailand. These border areas are the heart of the Golden Triangle where the hill people have been growing the opium poppy for centuries. Southern Laos and Cambodia are also becoming infected, not the least through rising infections in the refugee camps just across the Thai border.

The HIVs leaking north and south from Thailand tend to be two, increasingly distinct types. The southern variety is essentially the original HIV-1 on the genetic map, and displays the same moderate variability from one virus to another. The northern variety is recognizably different from the original HIV-1, and it is remarkably homogenous in its genetic sequences. It appears to be more prevalent in IV drug users, but no one knows when it arose or how it became so typical and distinct in the northern provinces. It will require a distinct form of vaccine, quite different from that which may eventually prove to be appropriate for the more prevalent and generally sexually transmitted southern variety.

Both the WHO and the American Army Medical Corps wish to test experimental vaccines, and the Royal Thai Army may cooperate owing to the high rate of infection among its soldiers and a growing realization that a true epidemic with catastrophic consequences is underway. In the same way that one minister noted "several Thai companies have realized that to keep up their profits it is vital [sic!] to keep their customers alive," so a growing number of senior army officers realize that they may not have many soldiers left to command if infection rates keep rising. Perhaps it was this realization that led the supreme commander of the Royal Thai Army to suggest that military regulations banning the conscription of HIV infected men should be amended. After all, commanding infected men is better than nothing, and as one cabinet member noted, "removal of the ban was aimed at finding an acceptable place in Thai society for AIDS victims." Whether the experimental vaccines will provide any degree of immunity to HIV is something that will take years of patient and carefully designed trials. Such trials require large numbers of people at potential risk divided into those who actually receive the vaccine and a control group who receive a placebo injection. Given the slow nature of the virus, it will take years before it can be established that the vaccine makes a significant difference.

What we have seen in Thailand is a catastrophe in the making, and in its eventual magnitude it may well exceed the terror of the killing fields of civil war in neighboring Cambodia. In Thailand a vibrant, yet gentle Buddhist culture is being torn apart and destroyed by the politico-medical equivalent of Pol Pot, except that the destruction is not the result of fanatical, Paris-trained marxists, but the socio-sexual fabric of the society itself. One cabinet minister, discussing tourism and the economic impact of the pandemic on Thailand, noted "It is the height of stupidity for a person to travel to another country on a suicide tour." But traveling is really not necessary. Suicide, like charity, begins at home.

9

America: leaks in the system

[AIDS requires] this country to slam the door on the wayward, warped sexual revolution that has ravaged this Nation.
Senator Jesse Helms, 133 Congressional Record *S14, p. 204*

Even if a man has not lived with dignity, our society's own self-respect demands that he be permitted to die with dignity.
Opinion in People v Williams, *New York State, Second Circuit, 1987*

To move from Africa and Thailand to the United States is to move to another world. We are now in the wealthiest of nations, with some of the finest medical and health institutions anywhere, a country with the capacity to stop the AIDS pandemic in its tracks, or at least slow it down so we have some sense of bringing it under control. Each state has its own public health department, funded at the state level, but augmented by federal money – after all, diseases do not respect state boundaries. At the federal level, there is a huge Department of Health, Education and Welfare, led by a member of the Cabinet, appointed by the President to provide direct and immediate access to the highest executive levels. Health committees in the House and Senate insure vigilance by Congress, manned (and occasionally womanned) by the representatives of the people. The National Institutes of Health are funded each year at levels exceeding the gross national products of many countries, and they are just what their names imply – institutions whose sole responsibility is the health of the nation. Finally, at Atlanta, the Centers for Disease Control are responsible not only for the records and research guarding the health of the nation, but also serve the United Nations and the WHO by coordinating disease control around the world. Once identified, no immunosuppressive virus could surely stand a chance against this array of wealth, knowledge and medical resources.

Unfortunately, by the end of 1991, the United States had roughly 60 percent of the world's cases of AIDS reported to the CDC and the WHO, using the new definitions that were supposed to go into effect in January 1992, but which were delayed by those same public health and medical bureaucracies. Even under the old definitions, the United States had nearly half of the world's reported cases, although we must remember that for many countries of Africa, Asia and Latin America the same reported figures are only gross and virtually meaningless underestimates. Nevertheless, and with the exception of homosexual vacation areas in the Caribbean (Turks and Caios Islands, French Guiana, Bermuda, Bahamas, and a string of other playground islands), the United States also has the highest rate in the Americas, a rate that is two to four times that of any other Western nation. Even these values are suspect and due for upgrading owing to chronic delays in reporting of up to three years. In 1992, for example, reported rates at the WHO and CDC were for 1989, since "reporting [was] generally incomplete."

Immediate questions are how did this happen and why? How, in a single decade, did things develop in this way? I am going to focus on that first, *how* question, and leave most of the second, the *why* question, to future historians. They will have to tease out the strands of a complex explanation involving the nature of the virus itself, the misuse of political power, holier-than-thou religious stances, impacted racial attitudes, individual sexual proclivities, and a sheer lack of will enervated by the languid neglect of selfrighteous leadership. At any one time and place all of these intertwined in various ways, informed by strident and often sickening statements of moral judgement on the right, and shrill posturing and babbling from the left, both sides exemplifying the certitude of rigidly held moral positions that left reasoned discussion with no place to go. In brief, it was democracy at work in a crisis. Nobody said democracies are good at everything: they take a long time to gear up for emergencies, and are usually late in recognizing a danger. Totalitarian states can act much more quickly, and often more effectively, which is part of their hideous appeal to people with little to lose. It is also perhaps one of the reasons why bureaucrats, even in democratic societies, have a tendency to close ranks and act in authoritarian ways that usually do more harm than good in the long run. As we examine how the HIV spread through the United States, remember that behind this question lies a complex mixture of democratic muddle.

We already know about the beginning of the epidemic, or at least as much as we are ever likely to know. The original tracing of sexual contacts (chapter 4) virtually pinned down the very earliest transmissions to homosexual encounters, and given the almost unbelievable levels of promiscuity of some highly mobile individuals (hundreds of different sexual partners each year), it comes as no surprise to realize that the seeding stage of the seventies included most homosexual groups in the major metropolitan areas. I want to pick up the account at the end of 1982, the year the HIV-1 was first identified, and a time when 1,485 cases of AIDS had already been recorded. In general, I want to show you the spread of HIV in a series of maps at the national scale, just like the sequence we examined in Ohio (chapter 6), using the accumulating totals over the first decade of the epidemic. If we wanted to, we could also follow the spread in terms of the changing *rates* of infection, rather than totals, numbers measuring what a physicist might call the acceleration of the epidemic, rather than just its velocity. I will actually start by using a rate map, simply to make a very clear and agonizing point. As it turns out, the general picture of how the virus diffused is not really that different no matter which way we look at it. Neither rates nor totals are entirely satisfactory by themselves. Rates tend to be misleading when the populations are small; totals tend to reflect increasingly the underlying distribution of the population – as we might expect. We shall try to work both ends toward the middle.

Before we start, I want to point out several things of a slightly technical nature. First, the simple and necessarily small-scale maps we are going to look at are based on the most detailed compilations ever made. Even by 1992, no maps like these had ever been constructed, even in the corridors of medical power at the CDC. The CDC has few epidemiologists with any experience whatsoever of spatial analysis, a professional background that could bring to bear a minimum of geographic insight into the epidemic. Even today the detailed geography of the epidemic is firmly hidden away from the public gaze by reporting figures only at the state level. We shall have to examine this question more carefully (chapter 11).

Second, while most of the rates and accumulating totals are based on counties (over 3,000 of them in the country as a whole), there are still a few states unwilling to provide even academic AIDS researchers with county values. Generally these are states with very small numbers of people with AIDS, and with correspondingly low rates, often with widely dispersed rural populations – the Wyomings, Montanas, North Dakotas, and so on. Others, like Florida and Wisconsin, with many of the most intense concentrations, insist that they know everything there is to know, that their teams of researchers are doing all the analysis they need, and they require no help whatsoever, thank you very much. Fortunately, we have totals for major metropolitan areas, as well as county values, and at the scale we are using here we can assign state totals with very little error.

Third, there is some under-reporting, even though legally AIDS and HIV infection are reportable diseases. More seriously, there are also reporting delays. This means that the latest maps may not show the full extent of the epidemic, and that computer predicted maps may well be closer to some unknown and unknowable reality than the official figures. We shall have to wait several years before this rather extraordinary claim can be tested, but I am prepared to bet on it right now.

Finally, slight errors here and there are unlikely to give anything but the most minute distortion to the overall picture, because in order to map these soaring rates and peaks on the AIDS "surface," we are going to use contour lines once again whose intervals will not go up in even steps, but in the ever-increasing steps of a geometric progression. In the examples here, each contour may be four to seven times the previous one. We have to show the change over time and space in this way because in a sense we want to have our cartographic cake and eat it. We want to be able to look at relatively small numbers when they first appear, but at the same time capture the sharp spikes of tens of thousands of people at the metropolitan epicenters. Just keep this in mind as your eye moves from the light to the dark shading on the maps.

By 1982, only New York and San Francisco were visible on the map of rates, two small pinpricks on the national map hardly showing up with rates of 29 and 16 per 100,000. But only two years later (figure 9.1), we can already see the tragedy beginning to unfold, although perhaps

Figure 9.1 Counties in the continental United States with AIDS rates of over 4 per 100,000 in 1984. Owing to the clustered nature of the distribution it is impossible to name every county. Those named, as well as others, experienced rapid increases in rates in subsequent years.

"explode" is a better description. Those 1,485 people converting to AIDS in the very first years of the epidemic have been joined by 8,945 others (a total now of 10,430), and we can see the effects of hierarchical and spatially contagious diffusion right before our eyes. That tiny pinprick of New York City has soared to 131, the highest rate in the country at this time, and is now the driving epicenter for the entire region, spilling over into Connecticut, New Jersey, and the counties of lower New York State. Notice how, around that Manhattan core, the Bronx and Kings are already 35 and 23 respectively, just ahead of Essex and Hudson counties right across the Hudson River – certainly no barrier to the HIV. And a little further out Passaic and Union Counties in New Jersey, and Queens and Staten Island in New York, are already rising. Beyond them there is a clear ring of Rockland, Sullivan, Westchester and Nassau in New York, and Hunterdon, Middlesex, Bergen and Mercer in New Jersey. If ever there was a classic example of the geographer's distance decay effect, a wave of infection pushing out from a major regional epicenter, this is it. If we simply plot the rates of these counties against their distance from downtown Manhattan (figure 9.2), we can see how they fall into a smooth curve (solid line), and how this rises dramatically two years later (dotted line).

What is happening here? We have a classical example of wave-like diffusion from an intense regional epicenter. Underneath those bare but terrible curves we have to imagine millions of human movements and contacts, some of them constituting the sexual and IV drug conditions for HIV transmission. Each day, hundreds of thousands of people move in and out of Manhattan, as though sucked in and out by great bellows, and in the suburbs other movements and contacts extend those potential chains of transmission which attenuate with distance from the center. In these early years we can see the contagious waves easily; in later years things become a bit more complex as similar waves of infection move out from other epicenters on the eastern seaboard – Boston, Philadelphia, Baltimore, Washington, and so on. Like the ripples of several stones dropped into a pond, the individual waves eventually interact and start to blur the simple circular ripples from a single center.

Even so, these waves of contagion can still be picked out in later years, right into the 1990s. In 1986, for example, Washington, DC, a huge spider sitting in the middle of converging routeways, already had rates of 104 (figure 9.3). If we take a circle with a radius of 70 miles around the Washington Monument, and plot the rates of the surrounding counties and independent townships in Virginia and Maryland, that classic distance decay effect is still sharp (solid line). Even four years later, in 1990, and despite other waves moving out from nearby regional epicenters like Baltimore and Richmond, the same distance effect can

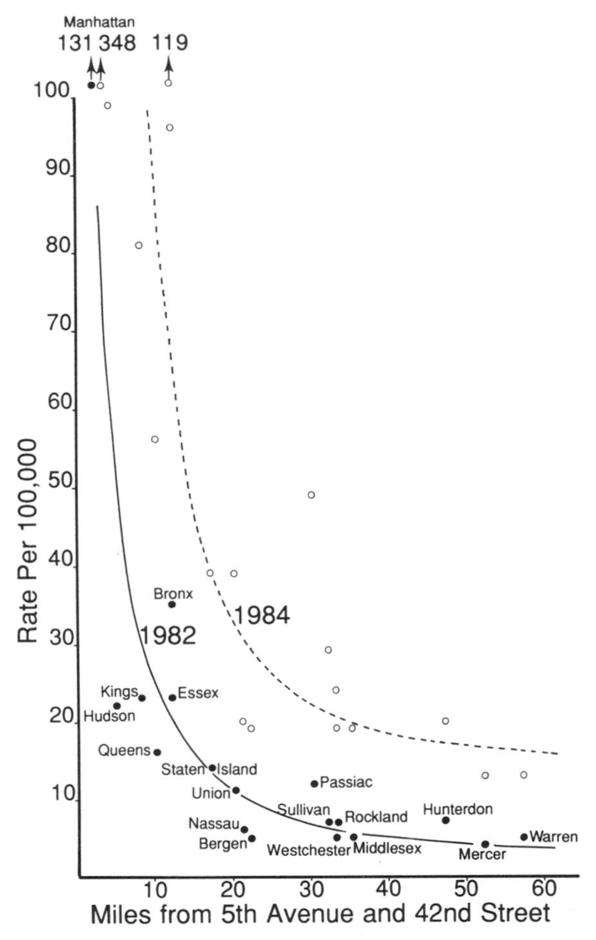

Figure 9.2 Rates of AIDS around Manhattan producing the geographer's classical distance decay effect from a regional epicenter. The effect in 1982 (solid dots and line) has become more intense two years later in 1984 (open dots and dashed line) as the wave of infection strengthens.

still be seen. It is now tragically higher as the wave of infection intensifies: Washington, Falls Church and Alexandria are not only off the graph, but off the page, with rates of 443, 206 and 192 respectively. The scatter around the 1990 curve (dashed line) is greater as other waves intrude, but it still easily discernible – for anyone who thinks like a geographer and has the eyes to see. Notice how counties like Dorchester, Talbert and Caroline, on the eastern shore of Chesapeake Bay in Maryland, are all "vacation counties" for well-off young men from Washington and Baltimore, weekend colonies producing much higher

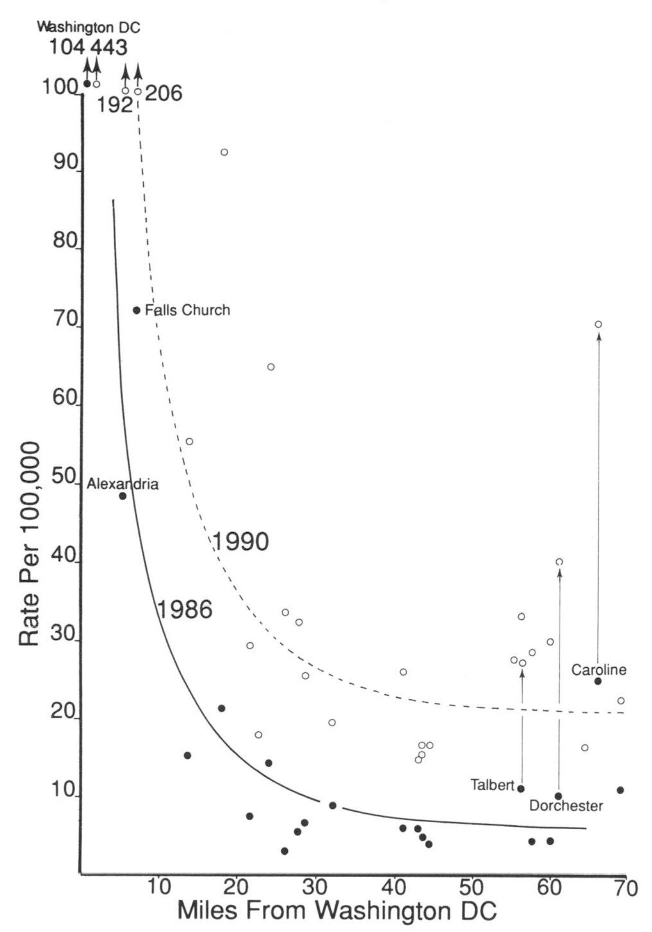

Figure 9.3 Rates of AIDS around the Washington, DC regional epicenter in 1986 (solid dots and line) and 1990 (open dots and dashed line). While the scatter of points increased in 1990 as a result of interference from other epicenters on the east coast, the distance decay effect is still easily seen.

rates of infection in the surrounding populations than we would expect from just distance away from Washington. A geographer would call these "positive residuals," and start to speculate why they occurred in a distinct cluster on the map.

But the point of these two distance decay graphs is that even by 1982, and right the way through until the 1990s, these thoroughly geographic effects were startlingly clear. If anyone had thought of looking at the *geography* of the epidemic, rather than simplistic numbers over time, we would not only have had a quite different and enormously valuable perspective on the explosive nature of the beast, but our thinking might have moved to something like predicting the next maps and using them as early warning systems. I think you will agree that maps have considerable persuasive power, and constitute graphical arguments that might have stirred the lethargy and benign neglect of even presidents and cabinet members. But no maps like these were ever made back then. After all, congressmen live in those districts, and we do not want to rock any boats in an election year.

And of course this was not just happening around New York and other north-eastern cities. Hierarchical diffusion had already made sure that Miami and its surrounding counties were going to be a disaster area, even though the AIDS researchers in Florida claimed they knew what they were doing. Over on the west coast the same thing was happening. The same distance decay effects were moving out of San Francisco and Los Angeles, from the former into Alemeda, Marin, Mendocino and Sonoma Counties, the HIV traveling straight up Route 101 paralleling the Russian River and its homosexual colonies strung out at intervals. Because what we are really looking at with these rates of AIDS is the movement of HIV along the same geographic corridors six or more years before. We know from studies here in California, and others from Texas, Florida and the eastern seaboard, that the dynamics of the AIDS epidemic are only a time-warped reflection of the same HIV diffusion. It is a slow plague to develop, but we can already see in 1984 how quickly it can move over time and space.

Even during these early years, simple maps are already loading themselves up with information, facts that are there before our eyes – if only we know how to read this graphic text – and assuming that someone had the geographic wit to construct it. Already the "recreational" counties favored by young and well-heeled homosexual men are showing up – Essex County just west of Lake Champlain, the Catskills in the southern tier of New York State, the Poconos in Pennsylvania, Summit County in the mountains west of Denver near Vail and other resort areas, and the beautiful islands between Victoria and Bellingham forming the county of San Juan in Washington. If we were to follow these and others over the next six years, we would see them become more and more prominent, shifting into higher categories on the map, some of them standing out like spikes from the surrounding areas. And even now, in this earliest and still faint human impression of HIV on the map, we can begin to see the emergence of that ribbon-like trace from Boston to Miami – the I-95 effect. By 1986, the pattern is already there along the entire eastern part of the country – Boston, New Haven, New York, Trenton, Philadelphia, Baltimore, Washington, Fredericksburg, Richmond, Savannah, Jacksonville, Miami – a string of infection

1,500 miles long whose early nodes reach out, coalesce, and intensify as the years go by. But nobody noticed it. Nobody said "Put your best, your most intense, educational efforts there, and there, and there." Nobody wanted to talk about nasty things like sex and condoms to young people in 1984, 1985, 1986 . . . and, after all, education, any sort of education, is a local matter.

Rate maps are not the only way of viewing the unfolding catastrophe. Rates are an abstract thing; by the time you have divided people by people, numbers by numbers, the people have canceled each other out and disappeared in a scientific computation. Carefully used, rates can reveal local intensities, and these are obviously important to know about, but the sheer human magnitude and cost of the epidemic escapes us. By 1982, those 1,485 people who had already converted and were dying of AIDS formed a distinct pattern on the map (figure 9.4). Already most of the major metropolitan areas of the country are counting their dead and dying. But notice the progression of shading and contour levels: each one is now 7.5 times the previous one, and once we reach 2,000 cases there is little we can do to capture the peaks of the epidemic that are literally "out of sight." As of writing (1992), over 18,000 in Manhattan (over 41,000 in New York), 15,000 in Los Angeles, 8,000 in Miami. If we could build a three-dimensional model of this "AIDS surface" in plaster, over a map the size of this page, and use a hundredth of an inch thickness to represent each person, we would have needle-like spikes over 15 feet high at these and other regional centers. Not even a computer can draw a three-dimensional map to capture important local details, and at the same time portray these needles of death.

On the west coast another alignment is already forming – Seattle, Portland, Eugene, San Francisco, Los Angeles, San Diego – matching the one we have already seen on the east coast. In between, the needle-like spikes of Chicago, Cleveland, Denver, and Atlanta are starting to rise, and we can see very clearly, two years later in 1984, how these have already become important regional epicenters for the outward diffusion into their surrounding umlands (figure 9.5). In a sense, and as far as the geographic appearance of AIDS is concerned, we are still in a "seeding" stage on the map. There is a scattered pattern of small numbers of cases across much of the country, some of them undoubtedly representing young men returning home to die. In 1984, and even in subsequent years, many of the apparently exact locations are best estimates, the assignment of state or large regional totals to major towns and metropolitan areas simply on the grounds of higher probability. In fact at this scale, and with these geometrically increasing categories, exact locations are not really the point.

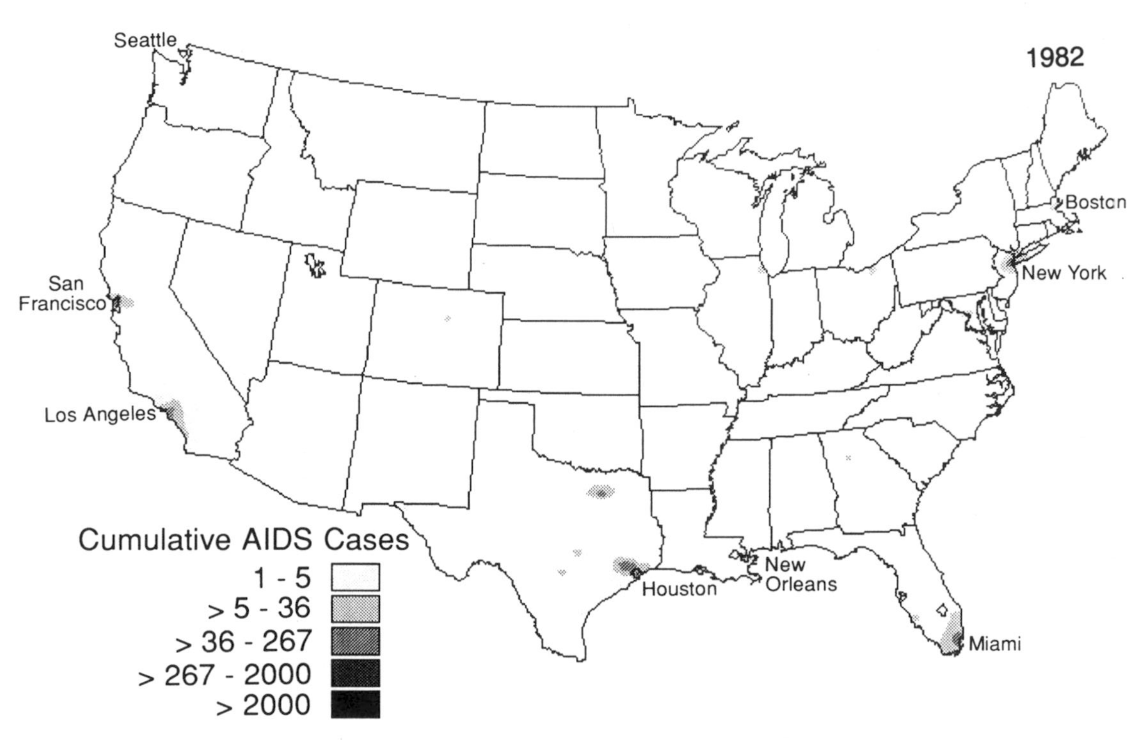

Figure 9.4 *The "AIDS surface" over the continental United States in 1982. On this map and others in the sequence (figures 9.5–9.8), be aware that each change in contour line and shading means an increase 7.5 times the previous value.*

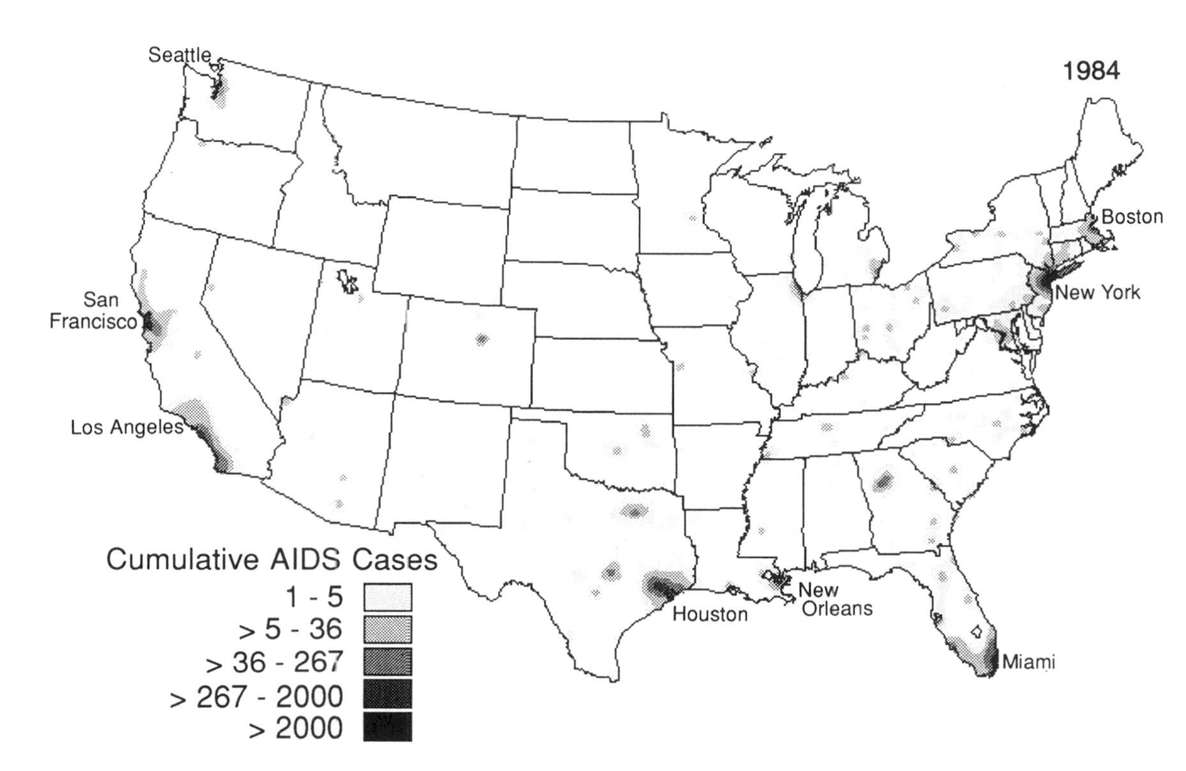

Figure 9.5 *The AIDS surface over the continental United States in 1984. Although still in what has been termed the "seeding stage," the geographical effects of hierarchical and spatially contagious diffusion are already quite clear.*

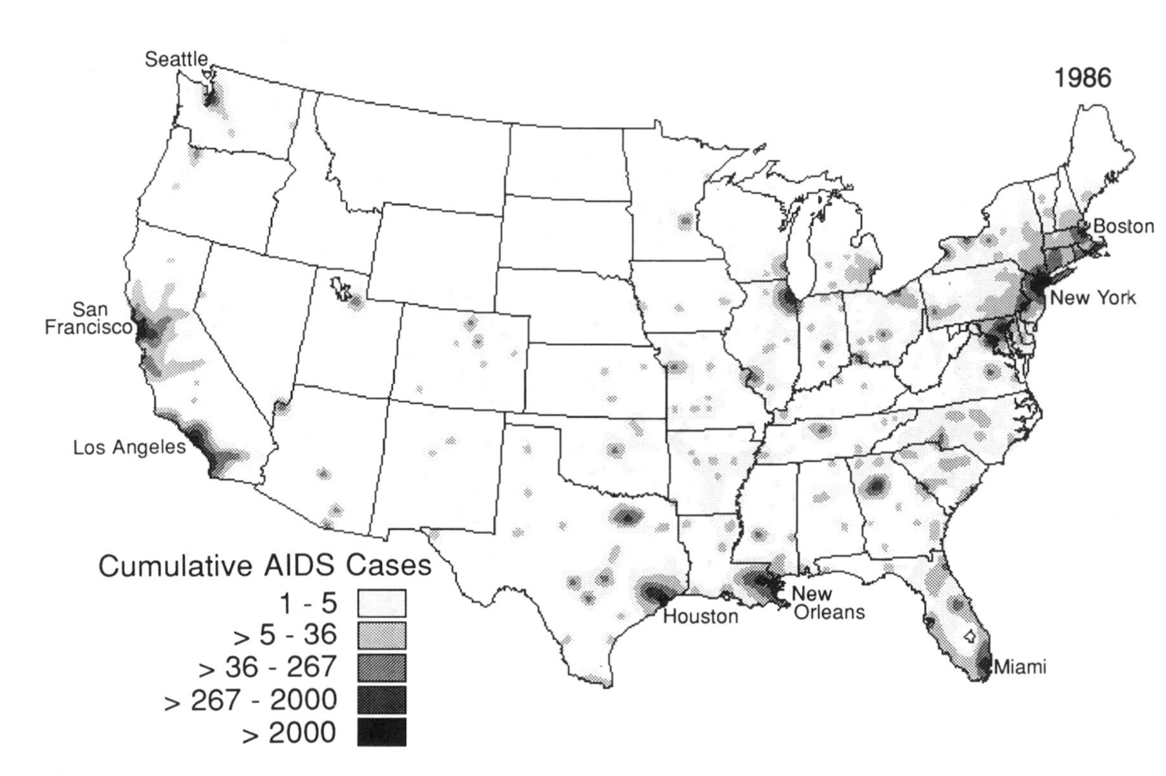

Figure 9.6 The AIDS surface over the continental United States in 1986. Major transport alignments and corridors of movement begin to be clearly visible.

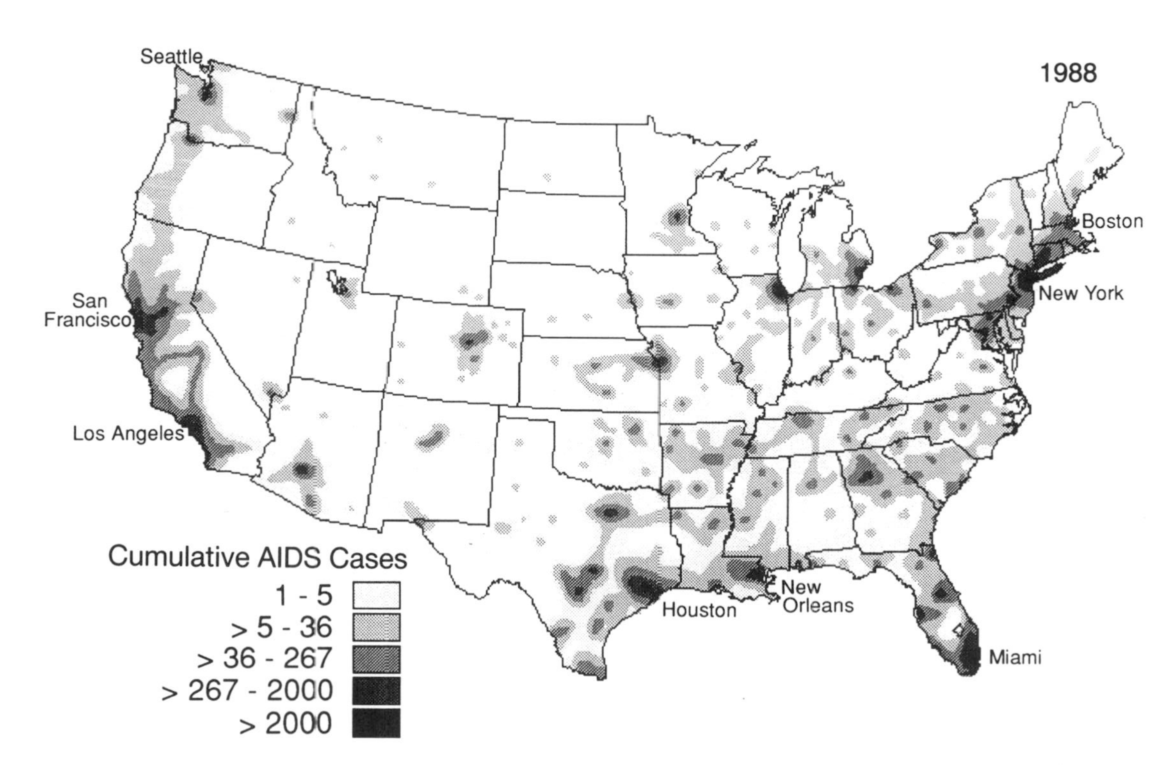

Figure 9.7 The AIDS surface over the continental United States in 1988. Major north-south alignments on both the east and west coast have intensified.

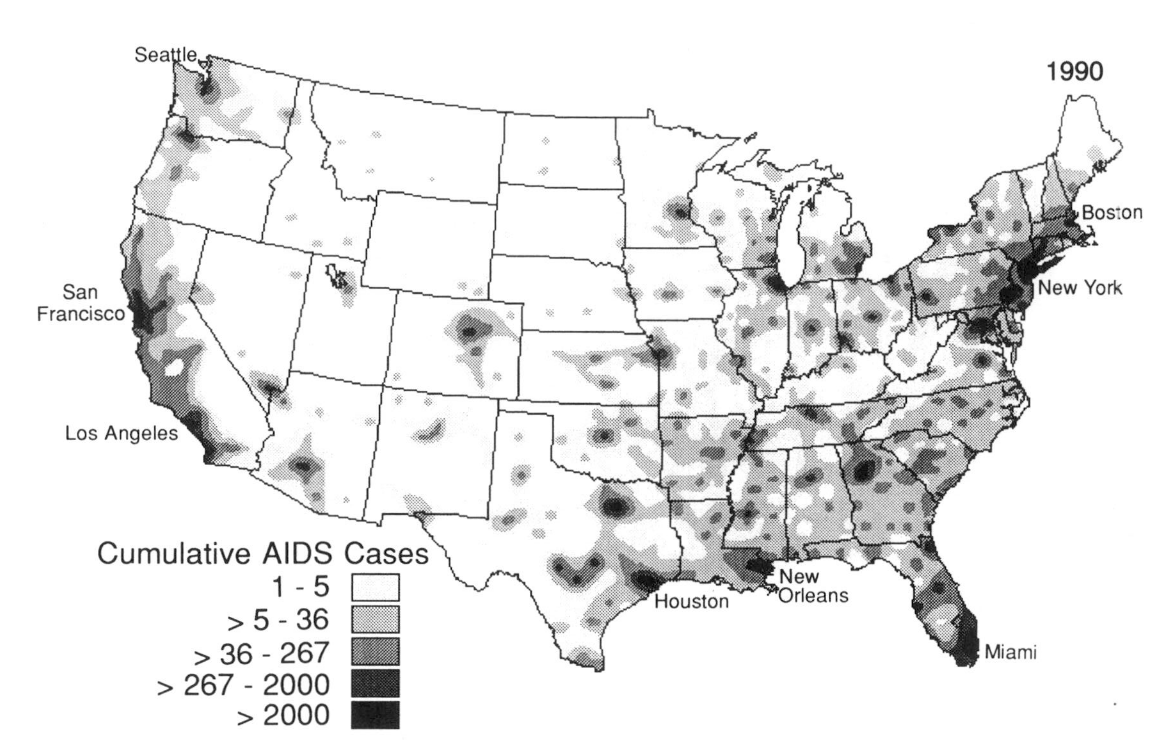

Figure 9.8 *The AIDS surface over the continental United States in 1990. Two years later, and using 1/100th of an inch (0.25 mm) to represent each person, a three-dimensional map would have sharp spikes of 13 feet (4.1 meters) high over such epicenters as New York and Los Angeles, and only slightly smaller ones over San Francisco, Miami, Washington, DC, etc.*

From sources inadvertently made available to me, I happen to know some of the patterns made by individuals returning home in the north central part of the country, but I have not used this confidential information, prefering to follow my general rule of assigning the small but growing numbers to the major towns. What other point would there be – except prurient interest?

By 1986 (figure 9.6), the epidemic had exploded to about 50,000. I say "about" because there is no point in trying to be more precise. Figures reported at the time by the CDC tended to underestimate actual values, due to delays, under-reporting, and still changing definitions as more was being learnt about the virus and its symptoms. The totals I shall use were compiled from individual state lists, some of which now contain corrections, late reporting, and so on. At this scale, and with numbers like these, obsessive precision is spurious anyway. On the eastern seaboard we now have a continuous swath from Maine to Florida, generally following I-95 with thickening nodes of infection along its entire course. Other major national thoroughfares are also starting to show: the old Hudson–Mohawk corridor via Albany, Syracuse, Rochester, Buffalo, and onto Cleveland and Chicago, reinforced by the more southern Philadelphia–Pittsburgh alignment. In the midwest, I-55 from Chicago to St Louis starts to appear as another major channel. As we saw before in Ohio, there are no surprises here: HIV moves from person to person without an intervening host (save a drug needle), and main transportation alignments are built to serve the movement of people. Any epidemiologist or medically-trained person with the slightest geographical awareness should have been thinking along these lines by the mid-eighties using commonsense rather than pretentious, and largely useless, computer models. Geographers, and I include myself among them, were also negligent in not pointing to the obvious idea that an epidemic always spreads in space as well as time. Although, to be fair, it was impossible to get even county figures in those years so obsessive and unthinking were the attitudes of the medical bureaucrats. Even the maps we are looking at here, and the colored versions on the cover, had to be compiled from individual state compilations with hundreds of hours of labor.

By 1988, the United States had approximately 120,000 people with AIDS (figure 9.7), and we can see how even at this scale one map seems to follow "logically" from the earlier ones in the sequence. Many of the white non-AIDS areas on the previous map have "converted" to the lightest gray shading, indicating a few cases are now being reported. In a few highly rural areas, there may be AIDS-free counties too small to show at this scale. In any map generalization using contour lines to capture the huge differences between urban and rural areas there are "spread effects" from extrapolation, but generally we have a reasonably graphic rendering with only small errors. The west coast is now a solid alignment of infection, and it is almost as though the advancing waves of spatial contagion from San Francisco, Portland and Eugene are throwing out pools of infection ahead of them, pools that will grow and coalesce into a major alignment some years later. It is impossible not to think of a forest fire advancing steadily, but throwing out sparks ahead of the main burning to start local conflagrations. By this time the peaks at San Francisco, Los Angeles, Houston, Chicago, Miami, and New York are already in the thousands, and from now on we shall simply have to remember that we cannot really represent cases beyond 2,000. Recall our three-dimensional plaster model: in 1988 those peaks are already 5 or 6 feet high.

Finally, by 1990 (figure 9.8), and really the last reasonably reliable figures at the time of writing, we have a devastating picture of about 160,000 people slowly dying, mainly as a result of transmissions in the early eighties, many of them before we even knew there was such a thing as the HIV. Obviously, this is not an exact geographic picture of the extent of HIV infection ten years before, but we can legitimately take it as a rough approximation. Of course, it is true that people move in ten years, and that we are looking at a country whose young adults are a group in the population with the highest mobility in the world. But many human movements tend to balance out (which is why toll bridges only need to collect in one direction), and the net movements tend to appear as much slower and larger scale drifts – from the rust belt to the sun belt – over decades. Given roughly ten years between initial infection and conversion to the many opportunistic diseases of AIDS, we are probably looking at a very rough approximation of HIV infection 8–10 years before. But who, in 1982, would have predicted these numbers, let alone the map of their geography? We tend to see what we want to see, and frequently display an almost limitless capacity to ignore the unpleasant consequences coming at us out of the future.

Is this fair criticism of those charged with the health of a nation, and equally those providing the enormous sums of money needed to combat the virus and its human consequences? Or is it just another example of the all too easy complaining that is based on wisdom by hindsight? For the early years, probably a bit of both. For the later years, since 1986, it is not only fair criticism but a stinging indictment of arrogant professional blindness and unconscionable neglect. It is also, and remains to this day, a classic example of fiddling while Rome burns, only this time the instrument is a modern computer producing different numbers depending upon who is playing the tune. As we shall see later, there is an obsession with computer modeling to forecast the epidemic down the time line, while remaining oblivious to its geographic and social dimensions.

In fact, no fancy, multi-equationed computer model is required to forecast what is going to happen in the absence of a vaccine. We only have to imagine a large, say three foot by five foot, three-dimensional population map of the United States cast in plaster of paris, with its great spikes of urban population dominating their surrounding areas. If we submerge our three-dimensional map and "population surface" in a swimming pool we can slowly raise it and record the "dry population areas" above the water level, as though geological time has been speeded up and the country was emerging above sea level. What would we see over five time periods of successive upliftings? Something very close to the 1982–1990 sequences of maps, the high urban peaks showing first and growing with each successive time, their surrounding areas emerging next, just like the spatially contagious waves spreading from the regional epicenters, and then ridges of dry land linking major peaks, just as we saw a gradual filling in along the east and west coasts, and increasingly the whole eastern part of the country. HIV is carried by people to people, and to forecast the gross outlines of this terrible epidemic you do not need obsessive precision, huge sums of money to estimate homosexuals and drug users, and so on. Increasingly the HIV is spread by heterosexual transmission, and in the most vulnerable – and valuable – group of people in the country, the young people, the people who will be the future.

And nowhere is the virus spreading faster than in those soaring peaks, those deadly needles, of the major metropolitan areas, places where the educational and social safety nets have been pulled out from under the poor, the homeless and the despairing. It really does not matter which metropolitan area we take as we change our geographic lens and focus down to look at one of them in detail. The story in general will be the same, expect for specific characteristics and details of places and people. It is time we looked at the Bronx.

10

The Bronx: poverty, crack and HIV

. . . we don't think it right to implicitly endorse the idea of sexual activity among our teenagers.

Spokesman for the Catholic Archdiocese of New York

It seems surreal that we're even talking about this . . . We've lost a whole generation already.

Richard Haynes, Director of Health Outreach to Teens, New York City

In the United States the HIV is everywhere, oozing from the high peaks of infection in the urban epicenters into the surrounding commuter fields and rural umlands, closing up the last empty pockets of white, infection-free areas on the map. Even five or six years into the epidemic, no one, least of all the professional epidemiologists, thought it would look like this, because few people other than geographers had been taught to think in the spatial domain. With two or three marked exceptions, thinking was always trapped in the dimension of time. Epidemics certainly have a history, but we seem perpetually doomed to forecast the future by asking the essentially useless numerical "when," but never the useful "where," questions. In the United States, the first maps of the AIDS epidemic I can locate below the banal scale of the states are one or two of New York City in 1986 and 1987. The first *sequence* of maps at any reasonably fine level of geographic resolution were constructed and published by me and some colleagues in 1988 for Ohio, around the same time as an enlightened epidemiologist at Walter Reed Hospital was mapping HIV from the medical examinations required by military recruitment. In other words, it took five or six years to come up with even the simplest cartographic picture of the geographical dynamics of one of the deadliest epidemics we have ever known. Perhaps somewhere some epidemiologist made a map a year or so earlier, but then had not the faintest idea what to do with it except hang it on the wall (chapter 14).

As we saw in the last three chapters, it is the "where" questions of the geographer that open up the spatial dimensions of the epidemic to thinking. Like those everywhere else in the world, the maps of AIDS and HIV in the United States are spiky surfaces, but the soaring peaks of the urban epicenters driving the geographic diffusion of the virus coalesce in many places to form mountain ranges of human infection and death, especially along the eastern and western seaboards. We are going to focus on the Mt. Everest of AIDS, New York City, and look closely at one piece, one borough of the city – the Bronx. It is a dramatic change of scale; our "geographic lens" now has a magnification power roughly 100,000 times the one we used before to look at the country as a whole, and we are almost at the point where some of the chains of people making up the socio-graphic networks transmitting the HIV are coming into view. Although not quite: most of these human structures making up the actual backcloth of the epidemic will always remain hidden, even to those at the forefront of medical care with access to individual records. But the magnification we are using now is sufficient to give us a dramatically different and detailed view and perspective.

It is also a scale and view that brings home once more the idea that an epidemic, like science itself, is always a socially embedded phenomenon with all sorts of human ramifications beyond those normally considered by the medical and associated professions. On this terrible peak of HIV infection we shall see all too clearly how everything – politics, economics, and culture – is connected to everything else. The Bronx is not only humanly devastated, with whole communities of people shattered into sharp and often dangerously antagonistic fragments, but it is also physically devastated. Walking through large areas of the Bronx is like viewing old newsreels of bombed-out European cities after the Second World War. Some have even written about the "desertification of the Bronx." Yet until the late fifties and early sixties it was a part of a great metropolitan area with many proud and cohesive human communities, some richer than others, some very much richer than others, but most of them with their own dignity and sense of belonging.

But the 1960s culminated in America's "urban crisis," with the Watts riots of Los Angeles and the looting of downtown Detroit as two high-profile examples, both of which were distressingly symptomatic of much deeper structural problems in the larger society, problems that were still around in 1992. Discrimination took almost every form of which it was capable; against the poor, against the different, against opportunities in education, in jobs, in housing, in health care, and ultimately in dignity and hope. It showed its face in high school drop-outs, low infant birth weights, high unemployment, rising homicides, more suicides, abandoned but still head-of-household mothers, drug addiction in every form, and later in the trembling crack and cocaine babies, and the multiplying deaths from AIDS. It also showed itself in the physical landscape of burnt-out and abandoned buildings. To understand the HIV epidemic at this particular place we have to start with the social and political structures that fell apart and allowed it to happen. If ever there was a human geography of this epidemic – a juxtaposition, a fitting together, a correspondence of spatial patterns – this is it.

In the roughly fifty years between the end of the nineteenth century and the 1940s, the Great Reform Movement of New York City had virtually succeeded in providing basic sewer, water, electricity, housing, and fire and police services. No urban utopia had been achieved – there were still slums and overcrowding, and occasional breakdowns in services here and there – but most people living in the city had a sense of community, of "our place." Like many northern cities, New York was a mecca for the rural poor, a tide of movement gathering strength over a century, and only reversed in the past decade or two. Many were black from America's South, while later arrivals in the fifties were Hispanic migrants, mainly from Puerto Rico. All, almost without exception, were poor. This was nothing new: for over a century New York had been the arrival point for waves of poverty-stricken immigrants, some of whose children stayed to form part of the city's life, while others moved on, and some moved up. The story, even in its inevitably mythologized form, is well known.

But for most black and Hispanic immigrants there was not much chance of becoming a meaningful part of the city's economic and political life, even less chance of moving up, and often no place to move on to. In increasingly overcrowded and impacted ghettos, the stress – what we might call the sheer "demographic pressure" – built up and spilt over. One spill-over area in the search for low cost housing was the South Bronx, much of it already in decay, an area of dilapidated housing and old industrial properties, most of whose owners had long since moved out. In the early 1960s, the Relocation Bureau of the city helped many Puerto Ricans to find new homes here, and they in turn formed a political power base for a borough president who later went on to Congress. By 1967 the South Bronx had become one of the designated poverty areas of the city with pockets of poor people scattered throughout the area. Later, as the housing and social networks were destroyed, many of these pockets grew and coalesced, diffusing in a spatially contagious way to reshape the built landscape and human geography of the region.

But the time (1967), and the place (the Bronx), were not just a piece of the spatio-temporal fabric of America floating in a void. The Civil

Rights Movement of Martin Luther King had started to break the old political hegemonies of the South, and more and more civil rights workers began to scrutinize similarly impacted political organizations in the northern cities. The Democratic "machine," a well-oiled consortium of the five boroughs of the city, began to fray at the edges: school boards once tightly controlled by the patronage system began to loosen the ties that fettered them, and in 1969 the first Hispanic congressman of New York, with a strong base in the South Bronx, won a significant number of votes for mayor, dividing a normally solid vote for the machine. Only a year later, and across the river in New Jersey, a black American would become mayor by breaking a similarly powerful and dominantly white machine.

Such hubris could not go unpunished. If the upstart Bronx wanted to play that game, so be it. There are many ways to bring the recalcitrant to heel, and probably no better way than squeezing them financially, particularly when you have a President and his White House staff at the federal level who, *sotto voce*, but in a leaked memorandum they only found slightly embarrassing, advocated "benign neglect" for the "minority problem." The problem of desperately overcrowded and poor people, in housing so run down that it often approached Engels' nineteenth century descriptions of Manchester's slums, showed itself in many ways, not the least of which were numerous fires. Almost without exception, every responsible fire chief writing at the time recognized the fire danger in old buildings with faulty wiring, electrical overloads, a population of heavy smokers, and piles of inflammable trash everywhere. Several astute analysts and professional firemen saw fires as a symptom of impacted poverty, and predicted both the spatially contagious diffusion of these, as well as the masses of poor people who would be displaced and needing shelter. Their predictions were far closer to the mark than those of so-called social and behavioral "scientists," most of whom were looking the other way. Most reports of the fire chiefs were directed at the politicians of New York, men in power who knew perfectly well what the situation was. By the late sixties, work loads for firefighting companies had become so heavy that the union forced hearings under binding arbitration to compel the city to open twenty new fire companies in the worst hit areas. One of these was the South Bronx.

But poor people living in dilapidated and readily inflammable housing were often seen as little more than nuisances, particularly in areas that might become new industrial sites if only they could be cleared out. Both "objective" academics, as well as professional consultants at Rand, had noted how "The lack of and high cost of industrial land is the single most important deterrent to the Borough's economic development." Not only were these damned (I use the word advisedly) people politically uppity, but they were obstacles to economic development as well.

The response of the city, both at the time, and even more searingly in retrospect, was unbelievable: between 1972 and 1976, the administration of the City of New York undertook a program of "planned shrinkage," either disbanding or removing from the areas most at risk 50 firefighting units, while reducing the number of firefighters by 25 percent. In six years staff were reduced from 14,700 to 10,200, while firefighting hours increased 45 percent, most of them in the high risk areas. Middle class areas were relatively untouched: the discriminatory reduction in fire services only mirrored the deeper discrimination against the minority poor, most of it completed well before New York's fiscal crisis of 1975, although most of the politicians did a Pontius Pilate act and were quick to seize this as a reason for their more-in-sorrow-than-anger speeches. By 1978, a Task Force report excoriated the reductions and demanded 45 new fire companies. Four were eventually put in, all in politically important white areas. The ghettos were left to burn.

As for the people who had tried to make their lives in these areas, large numbers of them also became "burnt-out cases" both physically and psychologically. Many became more and more desperate, with rising rates of drug abuse and violent crime symptomatic of communities falling apart. Building fires became increasingly numerous; some of them accidental; some deliberately set by slum landlords, who wanted to recover from insurance what they could not get from the housing market; some set mischievously, particularly in abandoned buildings – often next door to others housing people. Like a small burning hole in a sheet of newspaper spreading to the edges, so the growing burnt-out region drove a wave of fleeing homeless people ahead of it.

We can actually summarize the last two or three tragic decades in a formal way with a phase diagram, something scientists use to help them visualize how a complex system changes over time by tracing the trajectory it leaves behind (figure 10.1). New York City actually moves through a "poverty-fire space," whose axes are overcrowded housing (vertical axis) and structural fires (horizontal axis). We can even think of this in epidemiological terms: the overcrowded buildings are the "susceptibles" or the units at risk, while the burning buildings are the "infecteds." By the 1960s, tens of thousands of poor people had been crammed into rat-infested, slum-landlord housing, and sure enough the buildings they lived in provided a huge "population at risk." Between 1960 and 1970 the number of fires nearly doubled, and

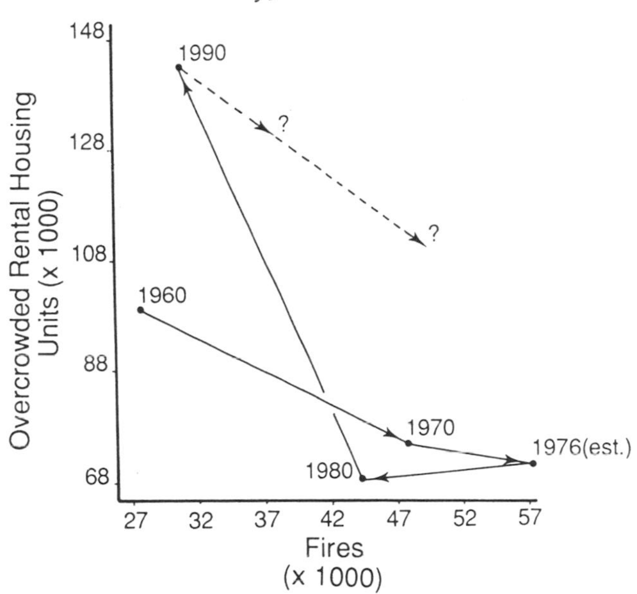

Figure 10.1 A phase diagram showing the trajectory of New York City in "poverty-fire space" between 1960 and 1990. In the early 1990s, there are indications that the cycle may repeat unless there are urgent and constructive interventions.

the people burnt out of their homes were displaced to other, often adjacent areas. This was the time when the fire chiefs saw clearly that miserable human conditions and fires went hand in hand, and under arbitration they got new fire companies opened. But by 1970, the policy of planned shrinkage went into effect, and the fires burnt more and more fiercely. By 1976 New York had 57,000 structural fires and nothing but reduced companies of desperately overworked men to contain them. Women could not help: they were considered too weak to do such "men's work," at least until 1986. As a result more people were displaced, and the overcrowded slums declined as the burnt-out areas spread to adjacent neighborhoods. By 1981 overcrowding was on the upswing again, and by 1990 the cycle looked as though it would start again. There is now plenty of space in the Bronx for new industries, but few healthy and well-educated people with hope in the future to work them.

What has this to do with the HIV epidemic? When stable backcloths of human relations in families, neighborhoods, and communities fall apart, then new, transient and deadly connections re-weave the fabric of daily life. They are the relations of shared needles and unprotected "sex for a penny" forming the structures for the HIV to exist and spread. In partially burnt-out areas the crack house moves in. In the Bronx you can plot the distance decay curves of the customers around what geographers call "central places." Heroin fixes and crack sales decline with distance from the source until it is worthwhile opening another house to serve the clientele of another locality farther away. Boundary disputes, otherwise known as turf wars, flourish where control is disputed. And where needles and sex form the connecting relations between people, the HIV flourishes. If you burn out whole areas, displacing people without hope, you also "shotgun the HIV all over the city," a phrase used by one prominent and dedicated analyst who worked at the Albert Einstein Medical School. This phrase cost him his job. The medical school receives substantial funds from the New York administration each year, and it is unwise for those who work there to point out that policies of discrimination masquerading as economic policies of planned shrinkage are responsible for accelerating the geographic diffusion of the HIV. Such a spattering of infected people over the city not only provides many new and more widely-dispersed point sources of infection, but it also makes educational intervention much more difficult, especially in fragmented communities where trust in the educational message of "The Man" has long since evaporated. The truth of this statement can be seen on any before-and-after map sequence, for example the geographic pattern of drug deaths before and after the burn-out (figure 10.2), but the truth is usually a casualty in politics. As the out-of-work analyst noted, "In New York they play hardball."

The result has been a succession of contagious waves; first the wave of burnt-out housing, followed by the wave of crack and heroin, and on the back of the crack wave rides the HIV. We can actually see it developing in space and time as a map sequence of deaths from AIDS, shown as a contoured "death surface" (figure 10.3). Based on counts in each of the 65 health areas, we can use these exploding figures as "spot heights" to see the geographical consequences of these prior political effects. The first four AIDS deaths appeared in the Bronx in 1982, but by the next year over half of the health areas reported people dying, most of them adjacent to the burnt-out region, the hollowed out human core of the Bronx. By 1984, the AIDS deaths were widespread, with the distinctive hollowed out pattern clearly established. Already the "death surface" is rising as a distinct north-south ridge in the west Bronx, with a rapidly rising plateau in the East, leaving a valley from which most of the people have been expelled by fire. Two years later, in 1986, these topographic features are even more prominent, particularly when we realize that the contours of death triple each time. A geography of death, like most geographies of human beings, produces a

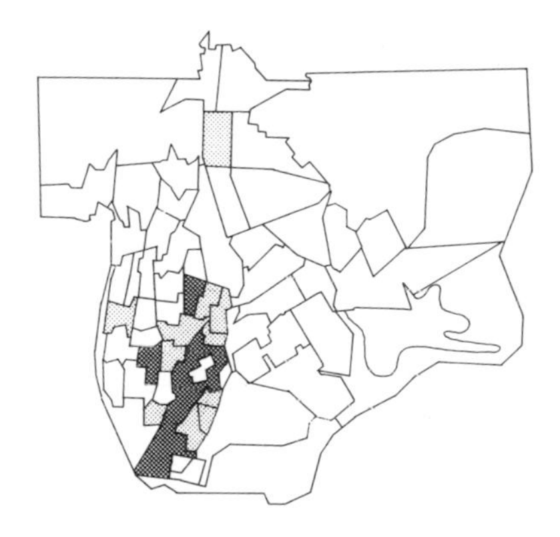

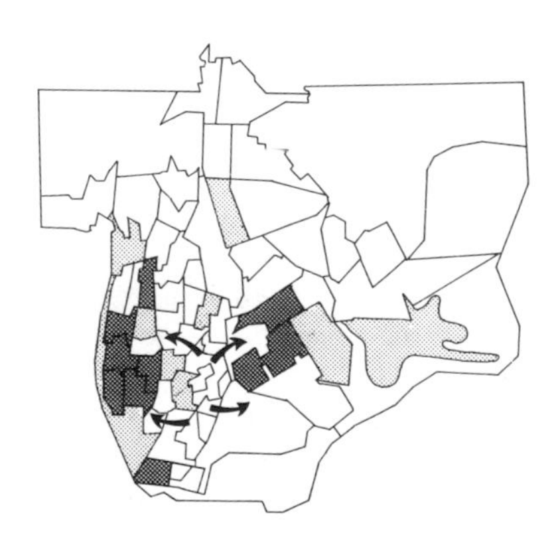

Figure 10.2 The patterns of drug-related deaths in the Bronx before (1970–3) and after (1978–82) the burn-out of the south-central part of the borough. Heavily and lightly shaded areas are the first and second highest-incidence subsets of nine health areas respectively.

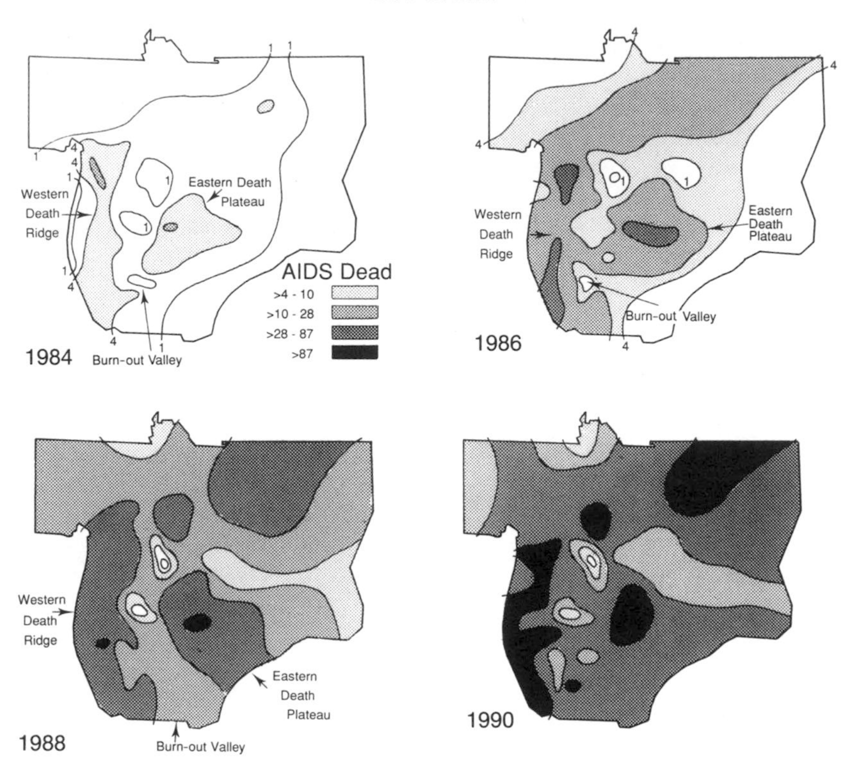

Figure 10.3 The rise of the AIDS "death surface" in the Bronx, 1984–90. By 1984 the hollowed out core can already be seen. Values between 1–4 deaths per health area are shown by marked contour lines. By 1986 the Western Death Ridge and Eastern Death Plateau have emerged clearly, only to intensify by 1988. The predicted map for 1990, showing widespread deaths from AIDS, is probably closer to some unknown truth than delayed and under-reported official figures unavailable at the time.

spiky surface. By 1988, AIDS deaths are literally everywhere and intensifying, with even sharper peaks emerging on either side of Burn-out Valley. Whether the politicians liked the truth or not, HIV was shotgunned over the city as a result of pulling the rug of fire and police services out from under the very people who needed them most.

The map for 1990 is a predicted one, based on geographic ways of using both spatial and temporal information to forecast what the epidemic will look like as it affects real people in real places, rather than being represented as abstract numbers floating around in a temporal void. We will look at these approaches later (chapter 14), but already this map raises a rather curious scientific dilemma that should give philosophers of science, as well as epidemiologists, something to chew on. I am going to assert that this predicted geographic distribution of AIDS deaths in the Bronx is actually closer, perhaps much closer to some "reality out there" (a reality known only to the gods) than the so-called official statistics for 1990 available two years later (1992) for the 65 health areas. Although based on public documents (death certificates), these aggregated figures by health area receive only limited circulation. There are also chronic delays in reporting, and some severe under-reporting, particularly in the major metropolitan areas of the United States – in other words, precisely on those peaks of the death surface where the epidemic is raging most fiercely. For example, in May–June of 1991, 75 percent of the AIDS cases reported in Washington, DC (another disaster area) had been diagnosed not a week or even a month earlier, but in previous *years*. It is becoming increasingly difficult to keep up with the virus and maintain a reasonably accurate and up-to-date picture of what is going on. Methods trying to provide reliable forecasts usually try to monitor and track what is going on as closely as possible, and so bring the latest information to bear. But it is no use tracking junk; all you end up with is another example of GIGO – garbage in–garbage out.

So this map is probably closer to the truth than anything the medical bureaucracy could produce even a year or two later, and what a geographical portrait of human catastrophe it is. The rising Western Ridge has now become a north-south wall of death overlooking Burn-out Valley, across which you can see inselbergs (isolated mountains) of dead people piling up to the north, east and south. These peaks are over the areas that were the housing refuges for people burnt out and displaced a decade or so before. The "benign neglect" of the sixties, the "planned shrinkage" of the seventies, and the systematic withdrawal of federal funds to support a safety net of hope and decency for "lazy welfare mothers" in the eighties, all combine and contribute to the epidemiological topography we see today. It is, after all, a slow virus.

This growing and uplifting topography is not unique to the Bronx; in place-specific and humanly particular ways the same story could be told in Harlem, Jamaica, Flatbush, Bay Ridge, and in equally terrible terms over the Hudson River in Newark and Paterson, New Jersey. In the poor pockets of one of the largest metropolitan regions on earth the virus is still moving. In New York City as a whole (the five boroughs) 41,000 people had developed AIDS by the end of 1992, and these were only the number officially reported at the time in a system known for its chronic delays. It was also a number reported under old definitions of what constituted actual conversion to AIDS. In 1992, a new guideline definition based on cell counts was meant to go into effect, a change that was expected to increase totals by 80 percent. For New York City it meant we were really looking at 74,000 people with AIDS, still based on chronically under-reported official counts. What the HIV rate is, what the future has in store, nobody really knows. By 1988, careful studies by people involved in drug treatment programs estimated anything from 60–80 percent of all IV drug users in the Bronx were seropositive, 20 percent of them women. The rates of seropositivity of women coming into Bronx hospitals for pre-natal care match those of Kampala, Uganda. By 1988, 5 percent of the children under 13 years old were also infected. The reactions of children are often as macabre as the old plague song "Ring-a-ring-a-roses"; a skinny kid will be pointed at by others who say, "Hey, I'll bet he's got the virus, man!" As one AIDS worker noted wearily, "It's just another plague visited on people who are plagued by other problems."

It is a plague rapidly seeping out of the bisexual and IV drug communities and into women, by heterosexual contact. By 1991, a quarter of all the cases in women were heterosexually transmitted and rising, 86 percent of them black and Hispanic, about equally divided. In the homeless of New York City the HIV rate hit 13 percent by the end of 1991, and was higher in homeless runaway teenagers. Quite how high no one is quite sure. As one homeless person in a temporary evening shelter said, "It's pretty hard to practice safe sex when you're living hand to mouth." And there are 200,000 people in line for the Human Resources Administration project apartments, some of them like the grandmother with six children bedded down temporarily (?) in Special Services for Children. It gets worse: whether in overcrowded slum housing, or overcrowded temporary housing, the women and children are at increasing risk. Especially the children. As early as 1975, astute social ecologists were noting the dramatic and accelerating rise of tuberculosis in areas of badly overcrowded housing. In these wretched human conditions the consumptive cough carries the TB from one person to another. Now, as adult immune systems go down, still more carriers are available to carry the bacillus to children. We shall look at this new synergism of plagues later, but its effect is particularly virulent on small children. Cases of TB diagnosed in children seem to reach a threshold in late 1991, and then suddenly tripled in a few months, 25 percent of the cases caused by a bacillus now highly resistant to any antibiotics we have available.

Speaking, or rather writing for myself, I have had to lead you through four of the most terrible narratives I have ever written. Numbers, percentages, maps and rates have been thrown at you, and it is easy to become overwhelmed and numb, perhaps almost as a protection of our own psyche, which, let us recall, was what the Greeks called the "spirit" of humanity – even if the psychologists have forgotten or are embarrassed. At whatever geographic scale – continental, national, regional or urban – the numbers are not just numbers, they are real men, women and children dying, and dying in the tens and hundreds of thousands. We must remember this always, but perhaps especially as we turn to examine the responses from the huge bureaucracies that have been brought to bear upon the pandemic, bureaucracies that also operate at distinct geographic scales from the global down to the local community.

11

The response: how many bureaucrats can dance on the head of a pin?

You waste funds that we need more than you do . . . A scientific approach does not produce the knowledge we need.
Reactions to the Managing AIDS Project, Eurosocial, *59–60, 1992, p. 12*

The public health authorities are not obliged to take chances.
Opinion in re McGee, *Supreme Court of the State of Kansas*

Life is full of crises: newspapers and news programs on television report scores of them as part of our daily and generally accepted diet of misery. Most of them disappear in a few days, forgotten by all except those whose lives have been permanently wrenched into new alignments. Many of them might have been avoided, but we seem to have a talent either for procrastination, hoping somehow that things will sort themselves out without too much effort on our part, or simply resigning ourselves to the seemingly inevitable. This syndrome of hoping for the best, while resigning ourselves to the worst, seems to be as characteristic of societies as a whole as it of the individuals who make them up. Perhaps tomorrow will not really be as bad as we think it is going to be.

But some crises are of such a magnitude, involve so many people, and have such immediate consequences that we are forced to respond. And actions are taken, always, and by definition, by individuals with the courage to take responsibility for them. When an earthquake, fire or flood arrives, you do not stand around discussing the pros and cons of maybe this or maybe that, or do you think it might possibly be a good idea? In the vernacular, you "get your act together" . . . and *act.*

Unfortunately, but inevitably, once the initial and often heroic actions are taken, larger institutions, both governmental and non-governmental come into play to manage a crisis. These are the bureaucracies, the huge, ponderous, faceless institutions with the capacity and power to bring the enormous resources of a society to bear on problems, but with an equal capacity and power to avoid action and bury responsibility in pillow-like procedures set in motion by decisions made by committees now long since disbanded. When a crises like the AIDS pandemic builds slowly, like the slow virus that is its cause, the possibilities for exercising and misusing power, while avoiding any responsibility for a decision, are enormous – often by relegating them to committees "to be formed sometime soon." The crisis of the AIDS pandemic is no exception, and the footdragging response of many of the major bureaucracies involved in the early years need not be told again. It was searingly detailed in the book *And The Band Played On*, one of the finest pieces of journalistic reporting of this century, published in 1987. The book caused outcries of anguished denials, from medical and health professionals to bureaucrats and politicians at all levels, but a high ranking analyst involved in AIDS research at the Centers for Disease Control admitted to me privately that "Except for a couple of things, he got most of the story right . . . unfortunately."

I want to pick up the story after those initial years, and look at a few, limited, but importantly geographic aspects of the way bureaucratic power, combined with a deadly combination of Establishment ignorance and arrogance, suppressed any consideration of the spatial dimensions of the epidemic, denying both the scientific community and the general public any knowledge, insight and real awareness of what was unfolding into a human tragedy. This personal account is not an "outlier" or solitary instance, even though accounts like these seldom get published, mainly because most people engaged in research are beholden to a very few sources of funding controlled by a relatively small number of people making up the reviewing panels. Not only do the circles of these reviewers overlap, but they invariably represent the established, currently popular, and taken-for-granted approaches. Philosophers and historians of science talk about "paradigms," meaning particular and accepted ways of looking at the world that dictate how research will proceed as "normal science." The notion of "paradigm" has been much discussed and criticized, but if ever a classical, almost textbook example were required, historians will find evidence for it here.

When bureaucratic institutions start to get involved in the management, steering, and occasionally the actual control of a crisis, there are usually four interrelated strategies that can be chosen. You can appoint committees, hold conferences, write reports, and throw money at it, often in a variety of combinations. These courses of action demonstrate to everyone that something is being done, and show that you and others making up the organization have a *raison d'être*, a reason for existing, within the larger federal, state or local structures. For this sort of busy, rushing around work you want "team players," the almost universal buzz phrase today, men, and occasionally women, who are good "committee people," who will work together in well-established ways, who know which side their bread is buttered on, and who will never seriously rock the boat with any ideas from "out in left field." In the course of appointing a committee or organizing a conference you might pick up a few with some original views and ideas, but you want to be careful here. By definition, those with original views and perspectives may not fit neatly into the established framework. They may (the ultimate bureaucratic nightmare), actually write minority reports when the fat and official Report of Recommendations is published, although by that time it will probably not matter very much because the crisis will either have passed, or grown in such proportions that what was recommended more than a year ago becomes essentially irrelevant now. The shelves of the National Academy of Sciences, and every other federally funded institution, are loaded with reports written in the innocuous style of committees, most of whose members want to make sure they will be asked to serve in the future.

Service on such committees and panels is not entirely altruistic. We are really talking about networks of power, particularly the power to allocate money, so insinuating yourself into such structures may pay off handsomely, particularly if you are a good committee person and resolutely steady the boat. The pressures to obtain external funding are mounting in the university, particularly where highly trained but distressingly uneducated engineers become administrators and treat the university as a machine to optimize the production of HPQUs – Human Performance Quality Units. The distinction between a for-profit research institute and a true university becomes increasingly blurred, a distinction that is dangerous to erase in a democratic society. Learning becomes more and more skewed towards fundable research, which invariably means normal science and Establishment research. As a result, "scholars" increasingly resemble marionettes jerking on the ends of strings held by government funding agencies. Too frequently these are administered by people who were incapable of imaginative and illuminating scholarly work themselves and so opted out for bureaucratic administration instead.

I want to illustrate a few of the ways the conference-committee-report-throw money strategies actually work themselves out to exclude any geographic perspective intruding into the established paradigm. To do this, I shall take you through four experiences of my own, starting with the White House conference in July, 1988, formed and supported by the Executive Office of the President. Its purpose, ostensibly, was to encourage "a national effort to model AIDS epidemiology" (the actual title of the final published report), a push for research no fair-minded person could quarrel with even if it was started seven years into the epidemic. No one ever said bureaucracies move quickly, especially when the White House knew it was a disease of homosexuals and drug addicts. Over 80 people around the nation and from five countries overseas were brought together from fields spanning the broad spectrum of the human, biological and physical sciences, and then divided into six working groups. Nobody quite knew what to do with a geographer, or quite why he had been invited, so I was assigned initially to the Data Collection and Quality group. After all, geographers presumably collect data of some sort – rivers and mountains and things like that? – so perhaps he will be useful there. When I noted that I had been modeling the diffusion of the epidemic with colleagues at two other universities for the past six months, and that the only reason I was attending the conference was to contribute what I could to the main theme, there was some puzzled surprise and I was switched to the Modeling Group.

The first meeting of the Modeling Group was something of a shock. In my naive way, and coming from the university world of ideas where it is usually possible to discuss a variety of perspectives and approaches, I thought we would start by tapping into the rich expertise represented by what was presumably the cream of the epidemiological modeling community. This was not to be: the first thing the "working group leader" (his official title) did was to place an overhead transparency on the screen with a series of differential equations on it. These are simply equations containing terms which are rates of change, in this case estimated rates of transmission of the virus. We shall look at these more closely in chapter 12, but without all the mathematical notations usually employed by researchers. Our "leader" was a mathematician who had made contributions to quantum mechanics, and he obviously thought the human world also worked like a piece of mechanism – although of a nineteenth century variety. Within a few minutes it became obvious that the "worlds" of most of the other participants worked that way too, particularly those from a huge national research laboratory largely instrumental in getting the conference supported in the first place. They were quite willing to admit that they needed the millions of potential funds to continue their modeling efforts strictly along these lines – the only directions their paradigm allowed them to see.

Those who have taken part in such efforts in other contexts will probably find nothing unusual or untoward here. Yet what is still surprising to some is the widespread acceptance of such machinations within communities claiming to be "scientific" and putting forward a public image of openness to free inquiry and new ideas. Too frequently the image is just that – a facade behind which the deliberations bear not the slightest trace of open scientific discussion, but resemble the smoke-filled rooms of a political machine worked by the power of a small caucus.

These "differential" approaches model mathematically an epidemic only and exclusively through time, churning out sequences of useless numbers whose magnitudes reflect in obvious and simplistic ways the "guesstimates" you started with in the first place. At one of the plenary sessions, held daily so all the working groups could meet together to "coordinate," one refreshingly forthright epidemiologist, who had spent the previous five years on the front line of care in a San Francisco hospital said, "I can predict just as well as you can with a French [plastic drawing] curve and a pencil." This was met with sympathetic laughter from other skeptics, but he was told, more in condescending sorrow than anger, that he did not understand the power of modern computerized models. Since these predominant differential approaches have failed utterly to illuminate the epidemic in any genuinely scientific sense, or provide any insight that might be of the slightest practical value in planning or education, we shall examine the failure of these efforts much more closely (in chapter 12). Our concern here is to see how the traditional response was pushed to the virtual exclusion of anything else.

To place this problem in a larger context, what we have is a particular instance of a much more general problem at the boundary between the scientific community and the general public, a public much more astute and well-educated than many believe, but nevertheless one often cowed by the mathematizing mystique of technical claims. In our modern world of almost unbearable technical complexity we seem to be obliged to leave more and more decisions to "them" – the experts and consultants who are attracted to public funds like moths to a candle flame. But too often the "experts" are obsequiously self-seeking and apparently unaware of the inadequacy of the taken-for-granted, unexamined, and terribly oversimplified assumptions that inform their technical fixes and recommendations. It is very difficult these days to find guards who will guard the guards, hardly a new problem as Juvenal knew 2000 years ago. But back to our conference.

In addition to flashing his differential equation, the group leader circulated multiple copies of a 29-page paper as a model of what the final report might look like four days later. The model report made a one page genuflection to statistical models, but then spent the next eight extolling the virtues of "evolutionary models," i.e. his differential equations. This was too much bulldozing even for the dedicated, so we were asked to split up into smaller groups to generate modifications and suggestions – if possible in written form so that paragraphs could be inserted. You have to close your eyes and imagine a dozen grown-up men and women rushing around and jockeying for position armed with word processors producing paragraphs supporting their pet approaches. After all, it had been hinted, $0.9 billion was going to be available for AIDS research next year.

I noted, in a tone of feigned puzzlement, the total absence of any geographic perspective whatsoever, and mentioned "spatial models." This took all but two of the group totally by surprise since the phrase obviously meant nothing to them. From his comments, the group leader obviously thought it meant rushing around from Alabama to Wyoming with a differential equation clutched in his hot little hand, but he invited me to "get it down on paper." This I did in two pages, pointing out the advantages of predicting not just numbers of people with AIDS, but where they might be. Several of the working group, including one who had worked with Russian colleagues, also thought this might have some merit, rather than being just an irrelevant distraction. However, it raised the question of data suitable for predicting the "where" as well as the "when," and with it the issue of confidentiality. Since this topic is essentially geographic in nature, and has been so grossly misused by the medical and related professions to withhold valuable data, we shall examine it very carefully later (chapter 13).

What became obvious over the next few days, as one draft replaced another in a medley of paragraph substitutions and sentence modifications, was that no spatial perspective was going to appear. None of the conventionally trained people had ever thought about it, it did not fit into any model that had been precast for the final recommendations, and the geography of the epidemic was going to be buried. I have each draft and modification timed and dated to this day: every time even a small paragraph appeared on the possibility of modeling in the geographic domain, it would disappear on the next round. Eventually even some of those open to new possibilities sensed that something was going on behind the scenes and made mild demurs. At the end of four days, the draft actually did mention "spatial transmission" – for whatever the phrase might have meant. It obviously meant very little even to those who wrote it, because it disappeared in the final published report where "spatial spread" appears once. All other references to "geographic" refer to doing conventional time modeling in specific places, the running around with a differential equation applying it here, there, and everywhere. Although there is a large literature on the geographic spread of disease, from fields as diverse as botany (plant diseases), biology (rabies), veterinary science (foot-and-mouth, poultry diseases, etc.) and geography (measles, mumps), our differential modelers were oblivious to it.

If something was going on behind the scenes of the intellectually parochial modeling group, something was also going on behind the scenes of the larger conference. For a variety of reasons, including even intellectual honesty, several groups, including the modelers, had obviously not done what had been expected of them. This was to mention prominently in their draft reports the absolutely top priority needed for a large national survey of 20,000 families, a survey even then in the advanced planning stage. This large and carefully designed random sample would not only take, and test anonymously, blood samples for HIV, but would attempt to record under the strictest conditions of confidentiality the sexual proclivities, behaviors and preferences of the American people. Only in this way, said the statisticians, could we estimate the degree of HIV infection, how it varied from group to group, and who might be passing it on to whom. The figures might also help us estimate the rates of transmission, so allowing us to calibrate our "guesstimates" and models. Whether these aims were justifiable and practical is something we shall look at later, but getting some sort of handle on the extent to which the virus had already infected the American population was certainly important. There was, of course, no geographic component or perspective to the survey.

No responsible statistician or behavioral scientist would launch out on a huge national survey like this without considerable and very careful preparation, including a series of small pilot studies to test the questions, responses and general feasibility of such an enormous survey asking for detailed and delicate personal information. Unfortunately, it was already running into trouble even as the conference was being held. One of the cities targeted was nearby Washington, DC, and the pilot study generated a storm of outrage from the local, predominantly black politicians. They charged that Washington had been picked out in a gross act of racial discrimination, that the people were being treated like guinea pigs, and so on and so forth. The fact that the charges were nonsense from any reasonable and scientific view was not of the slightest importance compared to the political heat and capital to be made out of them. If you want to test the feasibility of a national survey, it makes perfectly good sense to include as one of several pilot studies across the country a major metropolitan area where you know already there is a very high probability of HIV infection. The

HIV does not discriminate on the basis of the melanin content of the epidermis.

Back at the conference, the newspaper headlines produced a flurry of closed door conferences of consternation among the organizers from high federal and advisory positions. At the next plenary session, after several preliminary reports and recommendations by the working groups had failed to emphasize or even mention the proposed national survey, the iron political fist in the velvet scientific glove was allowed to show – just a bit. It became quite clear from "on high" that a major purpose of the entire conference was to promote the national survey with strong recommendations from the scientific community, and that final group reports would only be acceptable if they emphasized the crucial scientific necessity of the survey to their proposed research agendas. "After all," said the chairman from the private sector meaningfully, "we've all been around the racetrack." Indeed, most of the team players assembled by the conference had been around the racetrack – obviously several times. You have never seen such a flurry of activity in the next group sessions as fingers flew over word processors to alter sentences and inject new paragraphs. It was extraordinary how people suddenly realized how stupid they had been in not seeing before how absolutely crucial the national survey was going to be to their own scientific research. After all, $0.9 billion for research had been proposed for next year.

It was a curious experience, a display of individual agendas within group agendas within conference agendas that really had little to do with solving genuinely important scientific problems that by any stretch of the imagination were relevant to the AIDS epidemic. What was enlightening was watching, in an almost anthropological mode, the all too human relationships and petty power plays that constituted running around the racetrack. There were plenty of opportunities to watch, provided by the plenary sessions and the cafeteria-style meals where people from different disciplines were encouraged to mix and exchange views. Only occasionally did these appear to be shared. There was much skepticism on the part of the biologists, virologists and people in medicine towards modeling efforts, most of it thoroughly justified, and not a little jealousy that the focus of the conference was on the "modelers." One virologist said in essence "*We've* got $1 billion in research . . . what have you got?", but the jockeying for status went on all the time.

One of the most fascinating experiences, repeated on three different occasions, was watching the strange interplay of arrogance and anxiety displayed by the MDs at the conference, most of whom had opted for bureaucratic management rather than medical practice. You would gather your lunch or dinner tray, join a group of six or so that had formed out of the random flux, and start to talk about the epidemic. If there were MDs at the table they usually displayed a rather peculiar reticence, asking polite but non-committal questions that seemed strangely uninvolved. At last one of them could bear it no longer and said "Are you real doctors, or just PhDs?" Since a research virologist, an epidemiologist and a geographer, each with their pathetically inadequate PhDs, had obviously kept them on tenterhooks of professional snobbery for about 15 minutes without letting on, we told them earnestly and with eyes lowered that we were "only" PhDs. This appeared to give them considerable comfort, but after they left the three of us, bearing the burden of our intellectually inadequate qualifications, agreed it was too good a game not to be played to the full in future.

As for the report, it noted that the major findings of all the scurrying around were that mathematical modeling and statistical analysis of the AIDS epidemic were important, that the planned national surveys should go ahead "if they are found to be feasible," and that lots of money was needed to solve differential equations to produce numbers down the time line. The best research would come from collaborating and cross-disciplinary teams, and research funding priority would be given to these. But by that time the national survey was dead, and nothing more has been heard of it.

My second experience with bureaucratic response came five months later at a different level, when I lectured to the AIDS Surveillance Group at the WHO in Geneva on the approaches some colleagues and I had taken to modeling the geographic diffusion of the epidemic in Ohio, trying, already with some modest success, to predict not just the number of people with AIDS, but where they would be. In brief, I was showing them that there were a number of different ways to predict the next maps. It was a mixed professional audience, some MDs, a few epidemiologists, and in particular a group just back from Uganda, some medical people working on the international child immunization program. In the presentation, I pointed out the way in which the basic alignments of Ohio's major roads, railways and airlines appeared to shape and channel the diffusion of the epidemic, noting that this was hardly surprising since major lines of transportation are built to serve people, and it is people, after all, who carry the HIV. Nevertheless, this struck a distinct chord in those with fresh experience in Uganda, since it was already well-known that large numbers of truck drivers were infected, and even higher percentages of prostitutes and bar girls in all the stops from Ethiopia and Somalia to Mozambique and South Africa. Would it be possible to apply these geographic models to predict the next maps of East Africa?

My response really had to be another question: to predict with some confidence what the next maps would look like you needed geographically specific data about the people with HIV or AIDS . . . were such data available? It turned out that each month the WHO produced an updated data set based on a search of every published, and even unpublished, record available. The updated information was not actually compiled by the WHO, but a small group tracking worldwide HIV and AIDS in the Bureau of the Census in Washington. By this time they had established about 300 records on infection in Uganda and nearby Tanzania, Rwanda, Burundi and Zaïre. Would it be possible to put together a data set from these records to model the geographic diffusion of the epidemic? I said I would certainly review them and see what the possibilities might be.

What I found was evidence of almost unprecedented bureaucratic bumbling and a gross misuse of very scarce financial and medical resources. Without any sense of coordination, or that there might be a larger epidemiological picture to see, medical team after medical team had been funded to conduct surveys producing little if any fresh and useful information. In information theory, the fundamental idea of "information" is that it must have some surprise value. This is intuitively obvious and need not be dressed up in any mathematical or abstract way. If, on the basis of previous knowledge, you can make a pretty good guess about what you are going to find before you investigate, the information is not particularly valuable. It may confirm what you strongly suspected, but if you do not utter a "Wow! Hey, look at this! Who would have suspected it!", if you find what you thought you would find, most of the information is redundant. In other words, if you test the prostitutes of Bongo and find they are 74.2 percent HIV infected, is it really worth spending limited financial resources, and committing even scarcer medical resources (doctors, nurses, technicians, etc.) to test the prostitutes at Bingo and Bango down the road to discover they are 64.2 percent and 81.3 percent infected? To pose the question in its quite properly exasperated tone, "What the hell did you expect?" And does it matter for any real and practical purpose, as opposed to simply publishing your non-surprising and redundant findings in yet another medical journal, whether it is 60 or 70 or 80 percent? If it is 60 percent today, it will be 80 percent by the time you replicate your study a year later – again by misdirecting precious resources.

After about five years of cumulating medical studies, it turned out that you could establish reasonable estimates of HIV infection in the general population of Uganda at only two places and two different times. And you cannot do *any* modeling for forecasting, temporal or spatial, with two numbers with different (x, y, t) coordinates, i.e. at different places (x, y) and at different times (t). Millions of dollars had been spent by international bureaucracies in a totally thoughtless and uncoordinated way, money that had taken the few doctors and trained medical people from where they were really needed, and after five years we still had no overall picture of the way the HIV was spreading in the fundamental dimensions of human existence – space and time. This is not being wise with hindsight, a much too easy game to play, but a direct criticism of the national medical bureaucracies responsible for throwing money around East Africa and coming up with nothing of any real use after five years of effort. If you want to understand what is going on in a deadly epidemic you have got to monitor it in a systematic, well-planned and coordinated way. Then, slowly and carefully, you obtain data that people – medical educators, doctors, geographers, health administrators, even epidemiologists – can do something with. Otherwise you end with 74.2 percent of infected bar girls at Bongo.

I wrote a short report for the WHO, noting that the 300 records compiled so far contained essentially useless information, and urging that in future resources should be used in a careful and coordinated way to monitor the epidemic by generating data that could be used. This was not what they wanted to hear. When large bureaucracies ask the advice of others, particularly tame and "neutral" academics, what they really want is their preconceptions confirmed, and to be told that they are doing a great job. One may make a few minor and judicious suggestions here and there, of course, but nothing that can be construed as criticism. After all, you might not be asked for your supportive opinion again. As it was, I never received the courtesy of a reply, although I am sure that somebody, somewhere, now knows at considerable expense that the prostitutes of Bongo are 96.3 percent infected.

The third encounter with bureaucracy took place at the national level, and here we really trespass onto forbidden territory where charges of sour grapes can be leveled. I shall risk these for two reasons that I think are both pertinent and important. First, because this case study on research funding illustrates explicitly the difficulties of conducting research in the geographic domain when those controlling the purse strings live in a spaceless world, on the head of a pin like the proverbial dancing angels so problematic to medieval theologians. Second, because these inner workings are so seldom exposed to view because of the power relations inherent in the funding process. The sums of money are often large, and the pressures on academics to get external funding are often enormous and still rising – for the economics of substantial "indirect costs," for the prestige of the university, and for the freedom to inquire in time released from teaching. As a result, nobody wants or dares to rock the boat, to upset a major funding agency, or to ruffle the feathers of potential reviewers. For this reason, many research proposals, particularly in the human sciences, are as much exercises in deferential obsequiousness as they are statements of scientific programs of inquiry. There are whole courses given on the art of proposal writing, important parts of which deal with judicious ways of reviewing pertinent literature to flatter as many potential reviewers as possible. Nothing too original must be proposed, nor anything that might negate or contradict the accepted wisdom. One is permitted to indicate the possibility of making a modest advance, but always with the not so subtle indication that one is standing upon the shoulders of the giants who will probably be involved in the review process. Normal science needs team players.

And so a few months after the White House conference had extolled the merits of modeling and forecasting, and had strongly recommended interdisciplinary research, three colleagues and I approached the National Institutes of Health with a proposal for "spatial modeling of AIDS for educational intervention" – the actual title of the proposal – under an expedited review process that would only take nine months. Two of the colleagues at other universities were the world's experts on advanced approaches to geographic prediction, to predicting what the maps of people with AIDS would look like in the future, something we will look at in more detail later (chapter 14). The third colleague was one of the nation's top people and most effective researchers in the difficult area of adolescent sexuality and sexual education, a woman whose research record clearly showed she could work directly and effectively with teenagers and young adults, precisely the group of people we had identified as being the most vulnerable to HIV infection as it seeped out of the homosexual and IV drug communities that had taken the brunt of the epidemic in its first stages. Given the urgency of the health planning and educational tasks, we had no desire to engage in some academic exercise simply to pad our bibliographies. We had already made animated maps for television showing the AIDS epidemic jumping down the urban hierarchies of Ohio and Pennsylvania, and then seeping across the map like the now familiar wine stain on a tablecloth, and we had also seen the reactions of young people to these rather dramatic presentations.

People in health education talk about "cues to action," meaning things that really grab someone's attention and make a health risk immediate and personal, not just something out there that will happen to the other guy. Teenagers are notoriously difficult to reach, and famous for their immortality syndrome, the feeling that "it won't happen to me." We already had simple but direct evidence that animated map sequences on film, television, or computer screen, especially when they incorporated predictions of the future, had quite an effect on audiences. One film had been used for instruction at my university, the same sequences had been used at a school for young teenage girls, and one afternoon the cartographic laboratory of my department had hosted a group of black teenage kids visiting the university from Florida. They were tired at the end of the day, and getting more and more bored with being shown around meaningless scientific labs and libraries. But when we turned on the computer animation of AIDS spreading over Pennsylvania they were all attention. I wish I had thought of recording their reactions, particularly the young man who said with awe in his voice, "Wow, man! I never realized it was so close!" But how do you put this "cue to action" in a research proposal?

So we were proposing to model the epidemic in ways that might be immediately useful, predicting what AIDS was going to look like in space as well as time, helping people trying to plan future healthcare requirements, and using the results directly in educational intervention. That is why we called upon the expertise of our colleague in adolescent sexuality, to provide her with what we thought were powerful visual materials so she could test them as "cues to action" with the toughest audience you could find, but also the one most vulnerable to HIV infection. Two years later, at the International AIDS Conference in San Francisco, the "experts" would come to the same conclusion about the vulnerability of these young people.

But the review panel of NIH, loaded with MDs, conventional epidemiologists, and a "senior behavioral scientist" (his actual title), had never heard of the national call for interdisciplinary research, and they were distinctly unsympathetic to anything so useful as producing materials for educational intervention. You have to understand the close-knit nature of many reviewing panels, most of them formed from a relatively small pool of Establishment team players. Blinkered by the paradigms they were taught at graduate school years before, it is quite clear to them that there are only a few ways in which a scientific task can be approached. Most of them publish in esoteric journals of limited circulation, and would not dream of doing anything that could be construed as practical with their research.

As a result, the two parts of the proposal – geographic prediction and education – so carefully linked together were unacceptable and had to be torn apart. The geographic forecasting, employing mathematics and computers was OK, although it was clearly unconventional and did not genuflect to the sorts of things the "senior behavioral scientist" thought important. As for incorporating the geographic predictions in educational intervention, and testing their effectiveness as cues to action, this could conceivably come later. There was, after all, no hurry. Several years from now, when these rather strange and unconventional approaches had proven their worth, we might come back with another proposal to test whether these forecasts could be used effectively. In the meantime, the HIV would just have to spread through young people in the first years of sexual awakening and experimentation. Those making this decision called themselves The National Advisory Child Health and Human Development Council.

In light of these bureaucratic machinations of increasingly specialized scientists, it is difficult not to look back at epidemiological work in the nineteenth century with a sense of nostalgia and respect. Dr John Snow, for example, at a time when no one really knew the cause of cholera in London, made a map of its victims, saw them clustered around a public pump and put two and two together. He took the handle off the pump, and stopped the epidemic caused by infected water. In those days you did research to have some practical, roll-up-your-sleeves effect, not to publish in a journal to pad your academic *curriculum vitae*.

In fact, the research proposal was not turned down, because NIH seldom does turn down a proposal. Instead, it uses a particularly insidious bureaucratic device to insure that no responsibility can ever be directly pinned on an individual or even an identifiable group. It does this by invariably recommending approval, indicating in the initial covering letter that there will be a delay, and noting that no public announcement should be made until funding is actually assigned. This depends on a percentile grade and priority score, and if an approved proposal is too far down the list deep regret can always be expressed. What a pity . . . if *only* Congress had given the expedited AIDS research program more money, we would have loved to support your proposal approved by your scientific peers. In this way the responsibility is shifted from NIH and its long since disbanded committee to Congress. And Congress is even bigger and more remote than City Hall. There is no appeal, no possibility for further review. Our proposal was in the 32.4 percentile; the cutoff for funding was 16.4 percent.

"But what the reviewing committee was really trying to tell you," said the spokesman for the program at NIH, "was that you should revise and resubmit the mathematical modeling portion. If you note how it arises out of and extends more familiar modeling, I'm sure you'll get a sympathetic response." At first we wondered whether it was worth it. What was the use of *any* forecasting model if the forecasts were not going to be directed at something practical and useful? We already had 100,000 people with AIDS in the country, and estimates of a million or more HIV infected. Nevertheless, although I think now to our shame, we gritted our teeth and accepted the oblique but encouraging invitation to submit again the mathematical modeling we had proposed. Everything we had done so far, from gathering difficult and politically charged data sets, to writing complex computer programs, to producing animated films and tapes, had been accomplished under enormous stress in small interstices of academic lives filled with dozens of other responsibilities. Even to spring a little time free would be a gift from the gods. So we chopped off the educational intervention portion, mentioning it only as a sentence or two "for future research" (surely someone on the review panel would have a pragmatic streak in him somewhere?), gratefully accepted the reviewers' "suggestions," made all the right noises about the epidemiological tradition, while noting that we were standing on the shoulders of giants and this was just small extension and further contribution, and resubmitted the proposal.

Another, quite different review panel, again loaded with MDs and traditional epidemiologists, received the revised proposal and said there was "no demonstration . . . that the spatial diffusion models . . . will improve upon classic epidemiologic methods," i.e. models generating numbers "when" down the time horizon. The proposal was, of course, approved, but the percentile score was now 53.5. Ah, what a pity . . . that uncaring Congress has only given us funds down to 17.3 percent. The review report indicated that not a single member of the panel, all trained in conventional temporal ways, had understood a thing. For them AIDS existed in a spaceless world, and people with AIDS might just as well dance on the head of a pin. At that point we had had enough of hitting our heads on the brick wall of conventional epidemiology and gave up, slowly winding down the research over the next year, research that might have made a small but genuine difference in saving young lives.

It is difficult not to comment here on the extraordinarily closed nature of this reviewing community, contrasting it with other scientific areas where funds are readily available for modeling in the spatial domain. In agriculture, for example, the geographic spread of plant pathogens and diseases of farm animals is taken very seriously, not the least because modeling efforts are directed at important and highly practical intervention efforts. It is the same in the veterinary and wildlife management sciences: you do not record and model the diffusions of rabies just out of detached intellectual curiosity. You try to intervene, perhaps by dropping chicken heads inoculated with rabies vaccine in a *cordon sanitaire* around an infected area. Foxes, skunks and raccoons eat the chicken heads, become immune, and form a barrier to further infection. Farmers and vets are practical people and support practical research for intervention. But where are the Dr John Snows today when we need them most?

Fortunately, our work had not gone entirely unnoticed. A geographer prominent in the American Association for the Advancement of Science urged us to present this unique perspective on the epidemic in February, 1991, at the national meetings in Washington, DC. Perhaps it would be possible to get a hearing there; would I be willing to organize such a session? I accepted, and put together a series of presentations by colleagues who had gone through the NIH débâcle with me, together with an unsung doctoral student of mine who had done all the real work, and a new colleague who had found similar difficulties getting support for looking at the social and geographic aspects of HIV in the impacted ghetto areas of the Bronx. After ten years into the epidemic, this was the first scientific session anywhere devoted to modeling and forecasting in the geographic domain.

It was in some ways a dramatic presentation, both visually and in terms of letting some of the frustration show through. One of the things even geographers tend to forget is the sheer visual impact of sequences of maps showing something unfolding over time and space. There is nothing complex about these, you could give the maps and the numbers to high school students as an exercise in geographic visualization, but few people ever think of even doing this, let alone using the spatio-temporal information to forecast what the next maps will look like. As dramatically colored maps of the spread of the epidemic over the entire west coast were presented on the screen, you could hear the murmurs from even the traditional epidemiologists in the audience. One of the editors of *Science* was so shaken that he came up afterwards and asked if the journal could use them in an editorial report on the conference, a report that appeared only two weeks later in full color.

For other members of the audience, however, some of the presentations were not popular, particularly those that noted the puerility of conventional modeling in time. Some of the remarks in the papers presented were even hurtful to the finer feelings of epidemiologists and behavioral scientists in the audience who had forgotten that every disease and every behaving person has a geography as well as a history. As for using geographic forecasts for educational intervention, this was too much for the real behavioral scientists. "How do you know, what proof do you have, that these animated maps of diffusion will have a really significant effect upon the actual protective behavior of sexually active teenagers?", one behavioral scientist, probably an experimental psychologist, said scornfully. This was a comment we had met before, from the reviewers of our first NIH proposal. There we had hoped to demonstrate, with carefully designed before and after studies, that teenage attitudes to condom use, and their general awareness of the dangers of the HIV, would change after finally realizing "how close it was," rather than something remote, distant and out there.

But *attitudinal* change was not enough for the behaviorists; they wanted proof, hard scientific evidence significant at some conventional probability level, that these purported, but still untested cues to action would produce actual change in teenage protective behavior. This, it was strongly intimated, was what real human science was about, not this geographic speculation stuff. Attitudes, expressed as mere responses to before and after questions, were one thing, but how do you know real *behavioral* change has occurred, that the young couple in a state of early sexual experimentation and heightened teenage tumescence in the backseat of the car are really using a condom . . . eh? The absurdity of the question was clearly not obvious to many critics in the audience. After all, even behavioral scientists could not observe bedroom behavior, or go poking around lover's lanes with a flashlight and questionnaire asking young couples "Excuse me you two, are you actually using a condom, or have only your attitudes changed?" It was not impossible that in gathering such evidence the act of scientific observation itself could alter the behavior in the backseat, a problem not unknown even in the hard sciences. So I noted, with a completely straight face, that the only way evidence for real behavioral change could be marshaled was by inserting a latex-sensitive electronic chip into the penis of every young American male reaching puberty. This would transmit automatically via satellite to the computer at the Centers for Disease Control whether a condom had actually been rolled on, so allowing the behavioral scientists to report that 34.6 percent of teenage intercourse had employed latex condoms, up from 28.6 percent, a change not quite significant at the five percent level, and therefore scientifically unacceptable. "And if anyone here has a better idea," I said, "I wish they would let us know." After the laughter had died away, not a single suggestion was put forward.

I have taken you through these four, quite personal experiences of trying to enhance our geographic understanding of this terrible epidemic while working within a bureaucratic intellectual matrix that remains almost oblivious to the spatial dimensions of this deadly virus. In doing so I have lifted just a small corner of the rug, underneath which all sorts of professional jealousies and petty power maneuvers are being played out, usually at the taxpayers' expense. You have to realize that there is a vast and mobile army of underemployed researchers out there, many of them in the universities, many of them in the so-called human sciences. When a human crisis appears, and relatively large sums of money become available to investigate it, all sorts of people come out of hiding like hogs snuffling towards the trough at feeding time. Psychologists investigate the possibility that AIDS might be stressful; sociologists want to test rigorously the idea that the HIV might spread faster under type III socio-economic conditions, i.e. among poor people; and economists tot up the cost of the epidemic, dressing up their estimates with the latest econometric models when little more than the back of an envelope is required. Anthropologists appear as if by magic, each of them insisting that the place to study sexual behavior and HIV transmission is among "their people," usually a human preserve to which they have staked a possessive intellectual claim during their doctoral dissertation years.

And so the publications mount up, first hundreds, then thousands of them, rising in a steepening curve like the HIV itself. But not one in a thousand will have the slightest utility or make the faintest contribution in any genuinely practical situation. Few will illuminate anything of the slightest scientific or human significance. To put it bluntly, not one human life will be saved, not a single transmission will be stopped, by the millions of dollars spent in these areas that are always history by the time they appear in professional journals of limited circulation, read only by a limited number of obscurely specialist readers. Take a typical example: a research grant is given in June, 1988, and the research – finding suitable graduate students, contacting agencies, devising and pre-testing questions, etc. – gets going that fall. A year or eighteen months later some data has been gathered, and a follow-up extension requested to analyze it. Six months later the findings are written up and submitted to one of the hundreds of journals in the medical, epidemiological, behavioral, social, and related sciences. The review process takes another six months, and the editors suggest that some revisions might be appropriate – usually footnoting the obscure work of one of the reviewers. The revisions are made, and the paper accepted for publication, taking its place at the end of a waiting line. A year or so later the article appears in *Social Epidemiology and Sexual Behavior* (not its real name), one of the many journals, reports, bibliographies, and reviews that have sprung up over the past ten years. AIDS is a hot topic, and you can make money publishing almost anything about the epidemic. This means that in 1992 we shall know in great detail, and with all the paraphernalia associated with these "sciences," what was happening to a small group of people back in 1988. But by that time no one cares: we know nothing of any importance that we did not know already – the "prostitutes of Bongo" syndrome again – and nothing of the slightest practical use can be done with the findings.

Please do not misunderstand me: we are still at the beginning of a global epidemic that will take tens of millions of lives around the world, and we must try to understand what is going on to stop it in its tracks, and to alleviate the suffering any way we can. But simply throwing money at it is not going to work, any more than throwing money at cancer a quarter of a century ago worked. Research leading to understanding must be specifically and carefully targeted. What does this mean? It means huge, but still carefully thought through research programs in virology, immunology and pharmacology. If a "cure" is to be found, in the sense of a vaccine or medication, it is going to be found in these fields. Are there really any higher priorities? Then we must direct resources into caring for those already afflicted, which means extending healthcare in all sorts of ways, including enlarging hospitals, building hospices and housing, and training people. All these require planning, and all planning needs to know *where*. Finally, we must prevent transmissions, and prevent them increasingly in an area of human behavior that surges with great force through young people as they pass through puberty and into their sexual awakening. This means education. Any research not directed carefully and specifically at fundamental scientific knowledge about the virus and its human effects, at planning for decent and humane healthcare, and at preventative education is essentially asking the irrelevant "how many angels" question. Unfortunately, there are a lot of medieval theologians still out there, and it is time to look at one particular guild.

12

Time but no space: the failure of a paradigm

> Wherever power is concentrated today it tends to put modern technique to the service of droning banality, and oftentimes to that of the most blatant stupidity.
>
> *Jacques Derrida, in an interview with* Nouvelle Observateur

> . . . in the face of misery and suffering on a monumental scale, epidemic theory for its own sake is a luxury mankind can ill afford.
>
> *Nigel Bailey,* The Mathematical Theory of Infectious Diseases and Its Applications

To understand the way in which research on one aspect of the spread of AIDS has been so gravely distorted that it has been transformed into a parody of science, a pseudoscience, we shall have to enter a technical area that may appear to be so specialized that it has no place in a book written for a general audience. But this would be a mistake. Like many sciences using mathematics, traditional epidemiological modeling may only appear complex, when in fact most of the ideas and thinking on which it is based are actually quite straightforward and intuitively obvious. The problem is that simple and commonsense ideas may lead to rather formidable mathematical expressions couched in esoteric notations that seem arcane and incomprehensible to the uninitiated. This is a problem in many sciences, a barrier to more general understanding that is not confined to epidemiology. It is also quite true that some of the mathematical expressions may be difficult to solve in simple and direct ways, although we must always remember that this is a technical problem for the mathematician, and has little to do with thinking about what it all means. And meaning, like thinking, always has to be expressed and shared in words. Whatever the equations and symbols, if they are going to illuminate something worth knowing then someone has to interpret them, which means saying what they mean in words that others can understand. Otherwise they are literally meaningless.

As it turns out, epidemiologists never trespass beyond the gentle slopes leading to the foothills of modern mathematics, being content to use equations that might have challenged a mathematician of the mid-nineteenth century, but are routinely solved today on even quite small machines. The computer now lets us solve all sorts of equations that might have posed impossible challenges to very fine mathematicians a century ago. Some people actually like playing with equations on computers, particularly when they produce pretty pictures and graphics, and they tend to lose sight of the difference between doing mathematics that means something, and playing around in computerized sandboxes. We shall meet some of the latter in the pages ahead.

We should also remember that computers are only big and very fast adding machines doing the sort of arithmetic you use to keep a checkbook straight. For all the formidable expressions and equations, anything actually solved on a computer has been solved by simple arithmetic. So if you are one of those people who say "I never was any good at mathematics," then take heart. I promise you there is not one single equation in this chapter. Only words. Which, you will remember, were there at the beginning, and are the things which make us human.

But one caveat before we really begin, because I do not want to be misunderstood here. After all, some of my best friends are mathematicians, and I happen to disagree rather strongly with some educational psychologists at Harvard who say that anything can be made understandable to anyone at some level. There are many difficult areas in science, and some of them require highly specialized knowledge based on long years of learning to understand them. Remember the research virologist doing his best to explain things to scientifically trained colleagues (chapter 1). But traditional modeling in epidemiology is not one of them. Most of the ideas are intuitively obvious, even if their mathematical expressions look difficult. However, their simplicity does not necessarily make them unimportant. We are, after all, talking about ways of describing matters of life and death.

In choosing to model and express mathematically the AIDS pandemic, the epidemiologists have chosen a particular way of looking, a highly limited perspective that I am going to call the "differential paradigm." "Differential" means that they are concerned, and concerned exclusively, with *rates* of transmission, how quickly the HIV will pass from person to person, or from one group of people to another. And the fact that they focus on rates tells us immediately that their thinking is constrained to time – so many transmissions per day, week, month, year . . . whatever slices of time they choose to work with. If only we can estimate the rates of transmission, they say to themselves, then we can predict how many people will be infected *when*. Notice that the *where* has disappeared. We are back to our people with AIDS doing a dance of death on the head of a pin because the differential approach, in the form it takes, does not allow space and the geography of the epidemic to come to thinking. The distinguished German philosopher, Martin Heidegger, knew well how particular "theoretical" ways of looking allowed some things to appear and concealed others. He wrote:

> Theory makes secure at any given time a region of the real as its object area . . . [which] specifically maps out in advance the possibilities for the posing of questions.

And so we come to "paradigm" again. If you choose the differential paradigm, a framework and perspective that constrains your scientific vision to looking along the time horizon, then you are never going to think of asking the "where" questions. It was in this constraining sense that a distinguished virologist once called much of conventional scientific education "intellectual potty training." In fact, we could make a good case for the paradigm choosing the epidemiologists, rather than the epidemiologists exercising a free and thoughtful choice and choosing the paradigm. Even in the most distinguished universities and medical schools there is not one doctoral program in epidemiology that even hints at modeling in the geographic domain, in adding space to time. And we find this deplorable situation in a discipline and profession that can point historically to distinguished examples of the map being used as a crucial tool in the investigation of such diseases as cholera, malaria, dental caries, Kaposi's sarcoma, smallpox, measles, and many others. Even in modern science things once illuminated can go back into the darkness of concealment.

What does it mean to choose the "differential paradigm?" It means that you reach up to the shelf of "normal science" as you were trained to do as a graduate student, and take down your set of favorite differential equations. There is nothing conceptually difficult about these; they are only equations containing *rates* of transmission, and perhaps the best way of understanding them is to draw some pictures of what is going on, starting with the very simplest case (figure 12.1). Suppose we divide a population at risk into two parts, those infected at the beginning of the epidemic in the box on the left, and those who are still

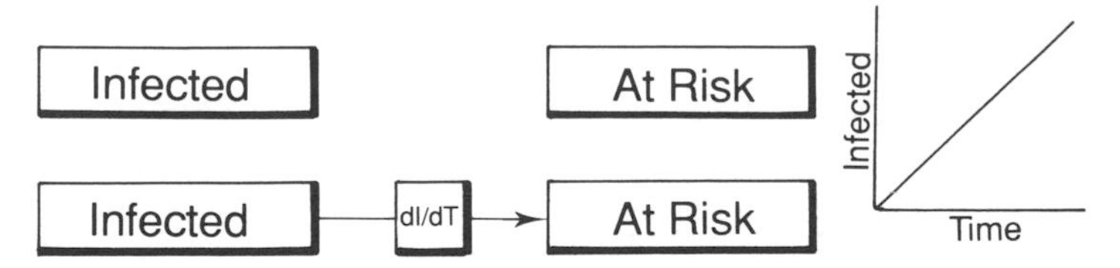

Figure 12.1 Two populations, one infected, the other still at risk. In the first case, the rate of transmission is zero because of quarantine. In the second, the two populations are connected by some constant rate of transmission dI/dT, i.e. so many new cases per unit of time. The connecting arrow symbolizes that the rate is unaffected by the number of infected people. In such a case, the number of infected people rises constantly (straight line) with time.

potentially at risk, the susceptibles, in the box on the right. If there is no connection between the two (essentially the equivalent of quarantining those infected), the epidemic is stopped dead in its tracks. Notice this is just another way of expressing what we looked at before (chapter 4). We are back with our people connected by sexual or IV drug relations making up the structures we called the "backcloth" needed for the existence and transmission of the HIV.

But the infecteds and susceptibles are not disconnected; there is some rate of transmission connecting the two boxes – so many of the susceptibles becoming infecteds per day, or per week, or per month . . . whatever slices of time we choose to take. A mathematician would say that there was a change of infecteds with a change in time, and using Δ as a symbol to represent "change," she would write $\Delta I/\Delta T$ or dI/dT. It is these "dee eyes by dee tees," or rates of transmission, that make up differential equations, and the more rates you have the more equations you need to work things out. We have a very simple example here, and if we assume, not very realistically, that the rate of transmission stays the same throughout the epidemic, then the numbers of infected people rise like the straight line on the little graph to the right.

As soon as we start to think about what is going on a little more carefully, a constant rate of transmission does not seem very realistic. At the beginning of an epidemic only a few infecteds are around, and they may contact only a few others to pass the HIV along. But as more people slowly become infected, there are more and more opportunities for transmission. This means that the rate of transmission, our dI/dT, also changes as the epidemic grows. We might reasonably expect that the rate itself depended in some way on the number of infected people, which we can show with a little feedback loop between the rate and the box of infecteds (figure 12.2). The more the infecteds the greater the rate of transmission, the greater the rate of transmission the more infecteds there are . . . and our epidemic no longer looks like a straight line but an exploding upward curve.

Figure 12.2 Two populations, infected and at risk, connected by a rate of transmission dI/dT. The connecting arrows now symbolize that the rate of transmission itself is affected by the number of infected people, i.e. the more infected people there are, the higher will be the rate of transmission. In this case, the number of infected people accelerates over time.

But as we think about things a bit more even this is not very realistic. To keep that exploding curve soaring upwards we would have to have a rate of transmission rising all the time, and an unlimited number of people at risk. In fact, we always have limited numbers of susceptibles, just as a raging forest fire only has a limited amount of fuel to keep itself going. The rate of transmission also depends on how many people are left to get the disease, and how often they are contacted. In a very densely settled area, say a tightly packed city slum, the contact rate would be very high, which is precisely why densely crowded slums are at such high risk in many epidemics. On the other hand, in a widely dispersed rural population the contacts would be much less frequent. Our transmission rate depends on the numbers infected, the numbers of susceptibles left, the contact rate, and so on. Notice that the contact rate in turn may well depend on the density of the population, which is a purely spatial or geographic matter. But having been "potty trained" exclusively in the time domain we will forget about that complication. In any event, we have enough complexity to see that our exploding curve of the epidemic must turn over and reach a limit (figure 12.3), and we are back at our now-familiar logistic curve (chapter 4).

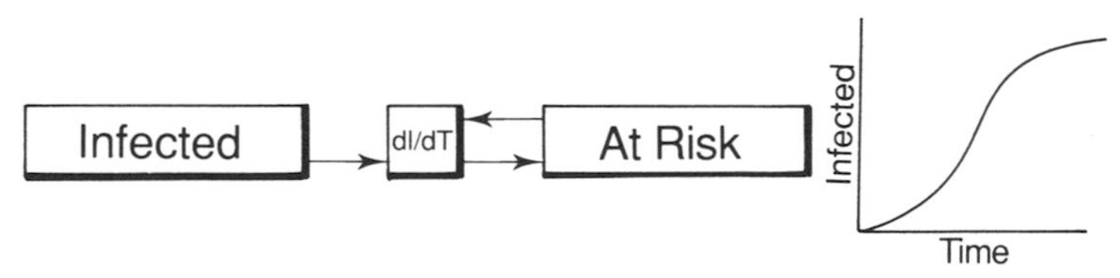

Figure 12.3 Now the connecting arrows symbolize that the rate of transmission, dI/dT, is affected both by the number of infected people and the number still at risk. The epidemic accelerates at first, then slows down, to produce the typical S-shaped or logistic curve.

Any equations we might put together along the lines of this very simple introduction to the differential paradigm could be solved quickly by hand, and they appear in any introductory text to epidemiology, physics, engineering, control theory, and lots of other scientific fields. But you do not get large research grants for simple things like this. You need to jazz it up, make it more complex, generate lots of equations, and solve them with computers. One way of doing this is to

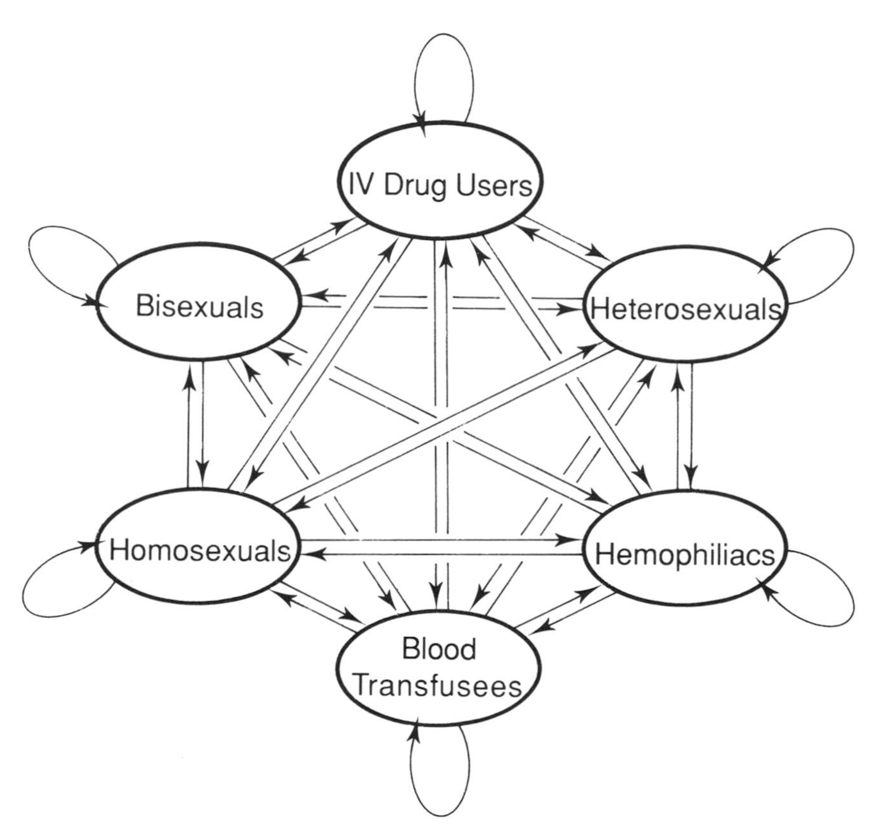

Figure 12.4 A population at risk divided into six subgroups, between which there are 36 possible rates of HIV transmission. If we divided these still further, by three ethnic, two age, two economic and two gender categories, we would have 20,736 transmission rates to estimate – rather too many to show here graphically.

divide your population into ever-finer and more numerous sub-populations, because this gives you more, bigger, and more difficult equations to solve, so at last you start looking like a real scientist. For example, and just to keep it simple, we might divide our population at risk into hemophiliacs, homosexuals, bisexuals, heterosexuals, IV drug users, and people who had received an HIV infected blood transfusion (figure 12.4).

Since the transmission rates between any two groups may not be the same, and people in each group can infect others in it, we already have 36 transmission rates to estimate in the middle of a raging epidemic in which the rates themselves are changing even as they are being estimated. But these rates of transmission may also vary depending on whether the people are black, white or Hispanic, or whether they are young or old, or whether they are rich or poor, or whether they are male or female . . . and so on. Ring all the changes on these still simple combinations, and we have 20,736 transmission rates to estimate. And, of course, lots of big differential equations to solve, needing bigger and faster computers, and, we regret to say, lots more money in research grants.

There is, of course, one small problem. In order to solve these equations we shall have to estimate the rate of transmission of HIV between, say, rich-old-Hispanic-bisexual males and poor-young-black-IV-drug-using females. This may pose some difficulties, but since we are theoreticians these are hardly our problem. Inhabiting the high and ethereal peaks of theoretical thought, we merely cast a lethargic eye over our minions in the behavioral sciences roaming the lower foothills below and tell them to get on with the grubby empirical work. After all, what else are they good for? And if they cannot come up with decent estimates, why this only demonstrates what we have said repeatedly – our models and theory have outrun the data. I promise you this is not a parody: the claim that theory has outrun the data has been made repeatedly by those clinging to the differential paradigm, and it appears in some of the most prestigious scientific journals in the world. The claim absolves the modelers and theoreticians of any responsibility, and allows them to pose as intellectual pioneers thwarted by the ineptitude of others.

So if we cannot get estimates of HIV transmission between even fairly gross aggregations of people at risk, we shall have to "guesstimate," usually a euphemism for blind guesses that are not totally "out of the ballpark." But now all sorts of pseudoscience pop out of Pandora's box. There is nothing to stop us putting all sorts of guesstimates in our equations, running our computers, and seeing what comes out the other end. Since everything is mechanically linked together, what comes out the other end reflects precisely the guesstimates that went in, and we reach the devastating conclusion that if transmission rates rise more people will become HIV infected. What other conclusion is possible? Alternatively, if we guess that as more people become infected and convert to AIDS so others will become more cautious and change their protective behavior, then fewer people will become infected later on. What other conclusion is possible?

So the banality of the whole approach has to be dressed up some more. We put in lots of guesses around an average guess, run our computer lots of times, and see what the spread or variation is in all the outcomes. Surprisingly, the range of outcomes reflects the range of inputs we put in. This is called "sensitivity analysis," and is both scientifically worthless and of no practical use whatsoever. It is scientifically worthless because we are only playing with equations in a computerized sandbox, and it is of no practical use for the simple and tautological reason that no one can do anything with the numbers. Except, of course, publish them in a learned scientific journal where they will promptly be forgotten even by the few who bothered to read them. Some modelers even generate their input guesstimates randomly, call their approach "Monte Carloing," and imply that they have moved from simple deterministic modeling to much more complex probabilistic modeling. But if they add up all the outputs from their random inputs and take an average, they usually end right back at the deterministic model they started with before tarting it up with random guesstimates.

In none of these approaches is there anything that can be even vaguely dignified by the name "science," a word we shall reserve for an honorable endeavor that seeks to illuminate some aspect of the world with human thinking. And when all the computing finally stops, you are still left with only a stream of numbers down the time line, numbers that few have any faith in because they merely reflect the last set of guesstimates that went in. But at that point nobody cares anyway. The AIDS modelers are like the demographers predicting populations in 10, 30 and 50 years time: demographer A says the numbers will go up; demographer B says yes but not quite as much; and demographer C says they will go down. Fifty years later someone is bound to be right, usually for wrong and totally unforeseeable reasons, but nobody checks or cares anyway. So it is with most AIDS modeling over even shorter time spans. Who in 1990 ever bothered to check what a five-year forecast was in 1985? And what would it matter? You could not do anything with the 1990 forecast in 1985, and there is nothing you can do with it now. At no point can you do anything with guesstimated "whens" when you do not know the "wheres."

But the obsession with playing with computerized equations continues. If you have lots of guesstimated transmission coefficients making up lots of equations, you can extract all the coefficients as a square table or matrix of numbers. Mathematicians would call this the

"Jacobian," and compute a number from this numerical array. If the number (technically an eigenvalue) is greater than one, the modelers proudly announce that the epidemic is increasing. The fact that this was patently obvious to everyone else in the country makes not a whit of difference, because computing the obvious from your guesses is much more "scientific."

How did we get here? How, in the name of the poor taxpayer, were very large sums of money devoted to these pseudoscientific "when" questions whose banal answers were useless to all concerned? In raising this, we are really asking about a specific example of the much larger and more complex issue of how paradigms, as scientific ways of looking, not only take hold, but even exert a vice-like grip on the scientific imagination, and command huge sums of public research money. If a paradigm carries along with it a great herd of scientific Gadarene swine, where is the swineherd to turn them around? Who guards the guards? Where is the piping voice of the child who says "But the emperor has no clothes on?" Examples of paradigmic shift, or "whoops! back to the drawing board," are legion in science, and one can make a good case for such corrections being at the heart of scientific inquiry. In the 1950s and 60s, cloud seeding was all the rage, and enormous sums of public money were poured into this research. Tens of millions of dollars later, some small child at the National Science Foundation had the bright idea that perhaps all the successes (when rain coincided with seeding) were published in the learned journals, while all the failures never got reported. After all, who reports a failure to the *Journal of Negative Results* even if it happens about half the time? And who hears anything about cloud seeding today?

Something similar has happened in the case of the "differential paradigm," and we shall have to go back to the late sixties and early seventies to understand the way in which a small coterie managed to get such a grip on thinking. It started with some pioneering research into ecological systems – how nature was all one interconnected web, with each part linked to and affected by many others. It was actually a very old idea, going back to the Count Buffon in the eighteenth century, a scientist who saw nature as an interconnected whole, and who was rather critical of Linnaeus chopping it all up into tidy little taxonomic boxes. The refurbishing of this old idea in light of modern knowledge coincided with the rise of the computer. It did not take long before the flow diagrams of the ecologists, diagrams showing the interconnections in natural systems and their flows of energy, began to be translated into differential equations. After all, the *rate* at which the sun warms the water, the *rate* at which energy is concentrated and amplified up an aquatic foodchain, and so on, are all important to understanding the complex dynamics of a natural system. And now we can not only write down the equations, but we can also solve them on our new computers.

Modeling ecological systems mathematically on computers took off like a streaking emperor. In the early and middle seventies it seemed that there was almost an unlimited cornucopia of research funds for this type of work. Of course, those rates, those transmission coefficients linking the ecological systems together, cannot always be estimated very well, but this is obviously a task for the field biologists, who now have a proper reason for measuring these things in the first place. It is not our fault if our theory has outrun the data. Sounds familiar? But then, in the late seventies and early eighties, some real ecologists – people who actually went out into salt marshes, or examined the delicate and complex natural interactions in the cold tundra – began to take a careful look at what the modelers had actually achieved in the way of scientific insight and useful knowledge. It was not very much. On a cloudy day the water warms more slowly than on a sunny day. Surprised? If the water warms quickly, the phytoplankton increase faster, the shrimps get fatter quicker, and so do the fish further up the line. What did you expect? And once again the piping voice of the irreverent child was heard in the corridors of the funding agencies. Funding declined precipitously in the late seventies and early eighties, just as the AIDS epidemic was picking up nicely. A quick shift from one funding agency to another, and off we go again.

But the emperor is still pretty cold. What we have here is one of many examples in science and engineering that have grown with the rise of our computing ability. A technique for solving a set of rather similar problems is found to be feasible given the enlarged computing ability we now have. There is nothing intrinsically wrong or inappropriate about such an advance in a limited area of problem solving, that step-by-step procedure usually described as an algorithm. We are able to do all sorts of things more effectively and efficiently as a result of these ways of obtaining solutions to tough and difficult problems. But in a way that must make the gods smile, latching onto a particular approach often makes us blind to alternatives, and very defensive in the face of any criticism. After all, people have often put their entire professional lives into a particular way of seeing and solving such problems, and the academic and research worlds are loaded with people who were potty trained as graduate students, and have clung fiercely to the methodological pot ever since. For many it is the only way they know how to be as scientists.

And so it is with our epidemiological modelers. The journals are flooded with one variation on the basic theme after another, all of them equally pretentious and unilluminating. We even have the ridiculous sight of anthropologists wandering around East Africa with their differential equations hoping to estimate transmission coefficients between sub-groups in the midst of a region where whole villages are being abandoned, hoping to "calibrate" their still purely temporal equations. The epitome, the ultimate folly of this approach came in a paper modeling the diffusion of the HIV in the whole of New York City with 34 differential equations churning out numbers down the time horizon. Geographically, of course, New York was homogenized and compressed to the head of a pin, simply because there was no need to consider any difference between the burnt-out Bronx and the trim lawn suburbs, or people in the packed tenement houses of a Harlem slum and the residents of apartments overlooking Central Park. After all, people are people, and since we are only playing computer games anyway, we can lump them together.

As can be imagined, the conclusions were carefully couched in the language of "scenarios," but nothing could really cover up their devastating banality. Computers can churn out scenarios by the hundreds, and they are simply the scientific equivalent of fairy stories. They are taken just about as seriously, because they only reflect the assumptions going into them. For example, if you have IV drug users (high transmissions) and heterosexual relations (lower transmissions) the number of infected people will rise rapidly at first (the needle is terribly efficient), level off slightly, and then rise again as slower sexual transmission takes over. Well, what did you expect? It was then proudly announced that the level of heterosexual transmission depended on the level of IV drug transmission. Put simply, if you have the perspicacity to give out clean needles, then there will be fewer HIV infected drug addicts to pass it along sexually. Well, what did you expect? As for the sensitivity analysis, the transmissions of HIV to children might be high or they might be low – it "all depends." Depends on what? On the huge number of combinatorial possibilities that can come up when you have 24 things you really cannot measure taking any one of 100 different values. Are you really surprised? Does this tell you something you did not know before?

For myself, and two colleagues who probably know more about the Bronx than any other people in the world, this was the final straw. We challenged, in an open letter, the major figures pushing these banal conclusions to an open debate on the merits of the differential paradigm, to be held under the auspices of any concerned institution prepared to provide a forum. One modeler said we had misunderstood the intentions of such modeling, although genuine scientific illumination was obviously not among them. The other said we had misunderstood his mathematics, a reply that did not endear him to one of my colleagues whose doctorate was in theoretical physics. Neither accepted the invitation to debate these questions in an open scientific forum.

Is there nothing of any worth here? No saving grace anywhere? Yes, one can point to one or two examples where the speed of the computer has allowed us to achieve some modest and carefully qualified insights. For example, what was going on in the early years of the epidemic is difficult to capture now simply because the HIV is a slow virus. Given a median time of ten years between first infection and now recognizable symptoms of AIDS, no one was recording anything in the seventies before those first appearances of clusters of rare pnuemonias in Los Angeles. Remember, we did not even know of the existence of retroviruses until 1977, and the HIV was not identified until 1982. So instead of forecasting, some useful work on *backcasting* has been done, to explore the sorts of very general conditions that might have been necessary in the seventies to produce the results we actually saw in the eighties. The research had no direct utility, it saved no lives, and it was essentially an exercise in historical epidemiology, but it opened up a chink of understanding we did not have before. The same approach has been used to monitor the epidemic at the national level, and forecast over a properly modest time span. The forecasts seem to diverge from the recorded values around 1987, and may be an indication of the way in which drug interventions are prolonging lives, although even these effects are doubtful once someone has actually converted to AIDS. Quite tragically the diverging lines may also be pointing towards a wave of deaths building up in the first half of the 1990s. The drugs may sometimes prolong, but they do not cure.

And that is about it, and at the risk of wearying repetition that is all it ever can be. If your thinking is fixed on time to the exclusion of space and society, you will never illuminate anything of the slightest use. So the question arises: why are the spatial dimensions of the epidemic, the geography of this terrible plague, totally ignored? Part of it, as we have seen, is because of the spatial blindness of traditional epidemiological training, and the grip of the purely temporal differential paradigm. But there is more to it than that. As soon as you start thinking about the "where" questions, as soon as you start putting things on maps and consider who the people are, the question of identifying a person has to arise – and quite properly so. Now if we were talking about children with measles, or young adults developing leukemia, or elderly people dying of chronic bronchitis, the matter of identification would probably not arise for the simple reason that few people would show any prurient interest. But with a virus mainly sexually or needle transmitted most societies take a rather different view. There is a social milieu of pruriency carrying all sorts of religious, cultural and economic overtones that most societies feel the unfortunate individual should be protected against. Many medical problems are seen as wholly private, and no one else's business. By thinking about the spatial "where," we have arrived at the ethically complex question of confidentiality. It is a question embedded in the geographic domain, which, as we have repeatedly seen, is precisely that arena of thinking where those who are medically trained have little or no experience. They do, however, have a great deal of power. It is time we examined this question carefully from the geographical perspective.

13

The geography in confidentiality

In matters of confidentiality, a man can trust only in his doctor and Allah, although as mortals we know nothing of their ways.

Ibrahim al Domice, eleventh century MS, Bukhara

The matter of confidentiality is an ethical one, arising out of an ethos that shapes people's views of what is right and wrong, fair and unfair, just or unjust. As a human construct it is not eternal and immutable, it is always socially negotiated, and it is frequently contested. What is acceptable in one society is unacceptable in another. What is seen to be fair and just, or as the Protestant prayer book puts it "meet and right," at one time may come to be seen as unfair and wrong at another. Ethical stances vary in space and time, they have geographies and histories, and as spaces shrink and as time speeds up it is hardly surprising that today philosophers are thinking about the problems of a global ethic appropriate to a small and shared planetary home. In the twin contexts of a mortal epidemic and a democratic society, the issue of confidentiality can produce enormous tensions, not the least because it brings into sharp relief the rights of the individual and the rights of the larger society. Does a wife have the right to know about her husband's HIV infection? Should emergency ambulance crews know when they are treating a bloody accident victim? Should testing be carried out? Anonymously? Should you tell people after they come up positive on repeated tests? Do you have the right *not* to know? Is it anybody else's business? Do my rights override yours? Yours mine? What is the right, the *proper* thing to do?

In most societies around the world a person's medical records and state of health are considered to be private, known only to a limited number of medical personnel who may be involved in treating the condition. This interpretation of confidentiality is generally regarded as normal and acceptable, but it may be raised to the level of the sacred – not too strong a word to describe the feelings some doctors have towards the bond of trust between them and their patients. So the medical profession has come to regard itself as the guardian of a sacred trust, and a conviction of sacred responsibility, a deep knowing that you are rooted in righteousness, is difficult to discuss, not the least because it may verge towards religious conviction where reasoned discussion becomes difficult if not impossible. Guardians of trust in a society are also very powerful since they manage information, and all power corrupts. Any trespass upon such power, even by the courts upholding a larger sense of social right, may be felt as a desecration and loss of power. We are on delicate and highly contested ground where feelings can run very high.

The ethical issues arise because the question of confidentiality is, in its essence, a geographic one. A medical record, an individual's dossier, normally records the time at which symptoms occurred, as well as the course of the illness – the times at which the patient got better or worse. Telling me about patient A, and the history of her illness along the dimension of time (t) tells me nothing about her – her identity is completely protected. But those same medical records also record a name and an address. Even if you erase the name, the address, that specification of geographic location (x,y), is enough for the prurient to break the confidentiality of the patient's identity. In the spatio-temporal specification (x,y,t) it is that (x,y) that has the capacity to break human trust with all the terrible ramifications that such a breach can bring. Notice the innocuous nature of the (t): it poses no problem, reveals no identity, and as we have seen, thinking about things along this dimension tells us hardly anything either. You can do temporal modeling to your heart's content because it discloses virtually nothing.

But as soon as you start asking questions about the geography of the epidemic, the "where" questions, then great anxiety is felt by the powerful guardians of medical trust, an anxiety heightened because few of them have any experience of thinking in the geographic domain where the question really lies. And the answer to high anxiety and a lack of experience is always "better safe than sorry" – always the response of the timid bureaucrat who backs away from personal responsibility and prefers to exercise power rather than reason. Oh yes, much better to be safe than sorry – especially when hospitals are profit-making and might lose patients, when politicians would rather not know, when the spread of HIV might be symptomatic of deeper social and economic conditions. Please do not misunderstand me: the question of patient confidentiality is one that any caring and compassionate person will treat with great thoughtfulness and respect. Most people will feel it is something that should be guarded carefully, and it is usually "most people" that make up an ethos out of which these questions of rightness arise.

But does the precise and confidentiality-breaking specification of a person's (x,y) mean we have to back away from any geographical analysis of the epidemic, when it is only the "where" questions that can provide truly informing inputs to health planning and educational intervention? From the response of many in the great medical bureaucracies around the world you would think the answer must be "yes." But it is a deadly yes, and deadly in two ways. First, it means we are unable to use the information about the epidemic in the spatial domain to save young lives and comfort the afflicted. In this sense it is, quite literally, deadly. But it is also deadly in the metaphorical sense: it constitutes a thoughtless exercise of bureaucratic power made more insidious by hiding behind what appears to be a cloak of impeccable ethical concern. Such a stance is almost unassailable and reduces most opposition to impotence. After all, those doctors, those nice, caring, ethically concerned doctors, know what they are doing. Surely?

Surely not. Especially those who have given up actual medical practice and have become the managers of the bureaucracies controlling the information. Many have honed the art of dissimulation to a fine edge. On three occasions now in the United States I have heard men high in the medical bureaucracy of the Centers for Disease Control say (the words were virtually identical over a span of 18 months), "Our policy [of releasing geographic data] is undergoing radical review. Get in touch with us, and we will see what we can do." Absolutely nothing happened then, and nothing has happened since. At issue was the release of data consisting of the simple numbers of people diagnosed with AIDS in each county, a data base compiled since the earliest years of the epidemic, but a data base that no one at CDC had the faintest idea of what to do with. In the first decade of the epidemic, no one at CDC had even thought about predicting what the next maps were going to look like, let alone the faintest idea of how you might use the valuable information in the geographic domain to plan ahead. Instead, data were released only on a state basis, a geographic scale that was analytically useless and scientifically absurd. After all, Rhode Island (State) fits into Texas about 250 times, and it is roughly the same size as many Texan counties.

But here we have arrived at the human and spatial question of the ways you can aggregate people into units of such a geographic scale that the individual person cannot be identified. Instinctively, a geographer would like to undertake an analysis at a detailed, "fine-grained" level, but is at the same time aware that if the units (say blocks in a town, or small cells of a square lattice overlying the map) were too small, then a single person or two might be identified by the pruriently curious, even if no other information about age, sex or ethnicity were provided. On the other hand, basic data about the geography of the epidemic could be aggregated into such huge units (for example, the states), that they would only constitute large and essentially useless geographic "smears" or spatial averages that would contain no useful information at all. You can imagine a map with a very fine-grained lattice of cells projected on it. If the cells are too small, someone might be identified; if they are too big, all the useful geographic information disappears. It is the "Goldilocks and the bears' porridge" problem. Not too hot, not too cold, but just right.

What does "just right" mean here, what scale of aggregation can give us a reasonably fine-grained look but still protect individual identities? The answer is that it depends on the human geography of the area. So you had better have geographers who are used to thinking about things geographic helping to inform the decisions about releasing data, not unthinking, anxious, well-meaning, but still powerful medical bureaucrats sitting on the data like dogs in the manger, not knowing what to do with it themselves, but refusing to let anyone else have it. Quite obviously, in a large and densely populated metropolitan area like New York, you should be able to release data on a relatively small geographic scale, say five or six city blocks, without the slightest possibility of disclosing the identity of a person; while in very sparsely settled rural areas, say the central parts of Montana and Wyoming, such a scale might immediately identify a single ranch house and home. You have to cut your geographic cloth to fit the situation, and for the sake of understanding the epidemic in space as well as time you better have geographic "tailors" wielding the scissors.

The resistance to releasing geographic data at reasonably fine but confidentiality-protecting scales has been extraordinary in some cases, while in others some thoughtful, but usually non-medical people have welcomed help in gaining another perspective. In Ohio, for example, an enlightened AIDS Surveillance Unit made data available on a county basis quite early, along with a number of other states, some of which actually published maps in their public reports indicating, in a crude and simple way, the geographic distribution of the epidemic. The state of Washington was among the leaders in this respect, probably because they had a geographer on their staff. The geographic quality of Ohio's data base, and the thoughtful cooperation of those who had taken such care in compiling it, made Ohio a valuable geographic laboratory for developing a series of computer techniques to forecast the next maps (chapter 14), and it was these that we incorporated into educational materials. In contrast, getting the same data sets from Pennsylvania was like pulling hen's teeth. Three explanatory letters and requests for information went unanswered, and it eventually took the intervention of the Lieutenant-Governor and the Chief Minority Whip of the House to pry a simple set of data from the virtual deathgrip of the then Secretary for Health – an MD turned medical bureaucrat. The data became the basis for the computer animated maps of Pennsylvania that produced the reaction from a young teenager "Wow man, I never realized it was so close!" Neither did we or anyone else: nobody had even thought of mapping the data before, let alone using it in any analytical way.

The video films and animated maps showing the spread of the epidemic were sent to all the state surveillance units and generated very mixed reactions. Washington immediately asked if sections of the tapes could be used for educational work, and of course we said "yes." Montana returned the video cassette with the covering letter date stamped and marked "received" – and that was it. One of the people working in AIDS surveillance, in a state that must remain nameless, knew from a previous conference contact that we were making computer animated maps and had begged to see them. The video cassette was returned by the chief of the unit, an MD again, with a curt note scribbled on our covering letter saying that his team would not be interested. I called the person who had requested it, who told me in a very fearful and embarrassed way that the tape had arrived, but that the chief had forbidden it to be shown even to members of the AIDS surveillance group. Yet at the same time, and across the country in Los Angeles, undergraduate students at Northridge were completing a "socially relevant" cartographic project under the enlightened guidance of a geographer, and with the full cooperation of the Los Angeles AIDS Surveillance Unit, producing beautiful, poster-sized maps at the census tract level, some of them only a few blocks in size. No one looking at the reds, greens and yellows of those detailed patterns could have possibly identified anyone. The following year, three map sequences showing the spread of the epidemic through the white, black and Hispanic populations were published, with each person represented by a dot placed at random within his or her census district. When the thousands of dots were located in the hundreds of census districts, you had a fine-grained picture of the course of the epidemic over the entire city (figure 13.1). Again, no one could possibly be identified.

The paranoid stance of the medical profession against the disclosure of any geographic picture of the epidemic is difficult to understand given these, and a number of other examples where confidentiality has been totally maintained. One epidemiologist had made a magnificent three-dimensional map of HIV infection in the United States based on county estimates, a map displayed at the International AIDS Conference in

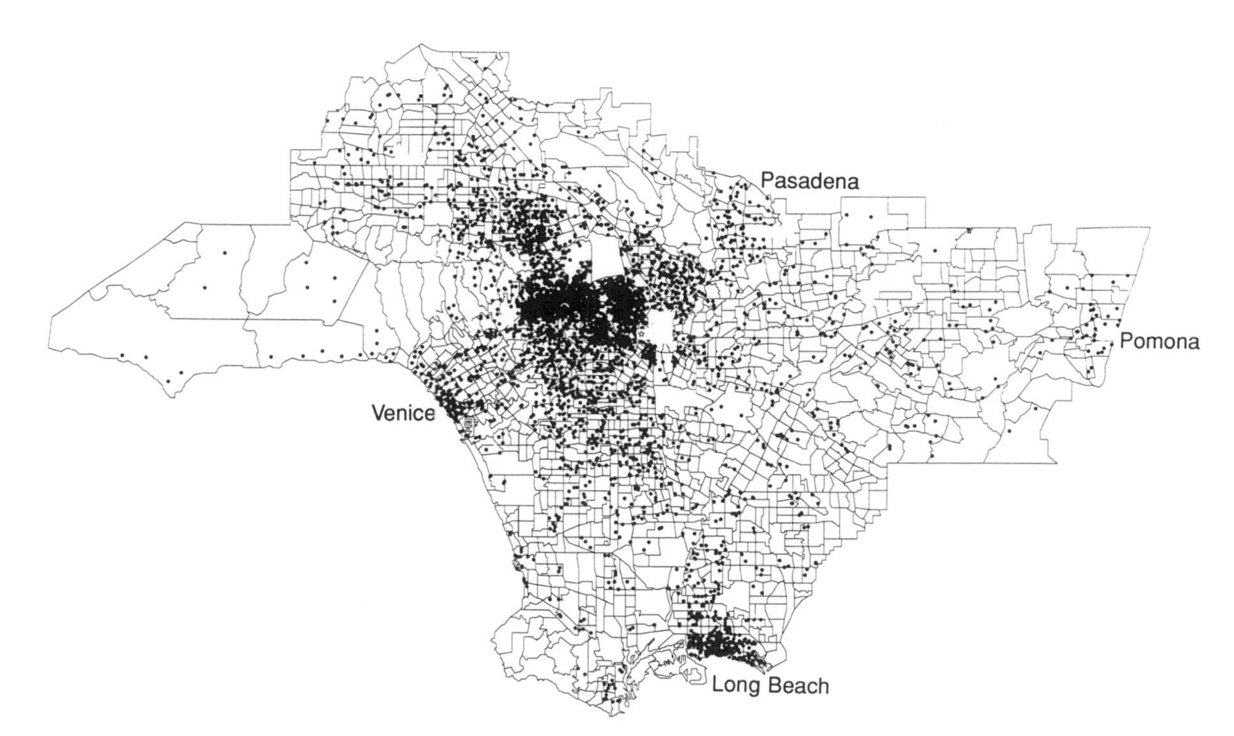

Figure 13.1 *The pattern of AIDS in Los Angeles by 1988, a map constructed by undergraduates at Northridge under the supervision of Dr William Bowen, and with the full cooperation of the AIDS Surveillance Unit of Los Angeles. The reader is challenged to identify a person with AIDS and so break the confidentiality rule.*

Stockholm. He tried to publish it as part of an article in one of America's most prestigious medical journals, but the editors, "upon the advice of reviewers," told him to remove it on the grounds "that our readers will not understand it." A third-grade child could have told you what those peaks and valleys of infection meant.

One of the problems of dealing with the confidentiality issue, apart from the blind obduracy of the medical profession, is that there has been little systematic research on the degree to which identity could be disclosed by making information available in small geographic units, especially when cross-referencing data are also provided. By cross-referencing, I mean indicating not just numbers of people, but whether they are male or female, in a particular age group, their degree of education, their ethnicity, and so on. For example, if you took even a relatively large area in certain parts of the western United States, and noted that a person with AIDS was a Navajo woman, between 35 and 40, with a PhD . . . there might be only one or two in that sparsely settled area that would fit such a description. This would be a clear disclosure, and properly forbidden. The problem is that to investigate the degree to which cross-referencing might disclose people in geographic areas of varying sizes you have to do research with real census materials, and in the course of doing research on the probabilities of disclosure you might disclose someone. It is Catch-22 situation – yet another one to hide behind if you are a bureaucrat – but also a situation generating lots of waffling, repetitive and "theoretical" review articles to bolster academic and research bibliographies.

Fortunately, some countries have much more open attitudes to census data, most of which are generally considered to be innocuous anyway. Quite frankly, I do not care if I am identified as a white, married male with three children, between 55–60, with an income of $50–100,000 a year. What did you expect, and are you really interested? And what professor makes over $100,000 a year – except in medical schools, of course. In Italy, for example, the computerized census records of over 3.5 million individuals were examined with names and addresses removed. As the authors of the investigation noted ". . . it can be argued that this is probably all that needs to be done to preserve the confidentiality of census data." Even when 40 cross-referencing categories were used to slice up all the records, the number of unique cases was very small, and using very conservative estimates to compute the actual probability of a person being identified, the chance came out to be 1 in 10^{-11}, or about one chance in 100 billion. But in the case of people with AIDS, or estimates of those with HIV infection, an enormous amount of geographical analysis can be accomplished with just the numbers alone, without any cross-referencing about sex, age, or ethnicity at all. Given the high degree of spatial segregation in the United States, it does not help very much to tell me that 96 percent of the cases in Harlem and the South Bronx are black or Hispanic. What else could they be? This "information" has no surprise value, and it is surprise value that sharpens up predictions. Similarly, telling me that most of the people with AIDS in Montana during the first decade of the epidemic were white males between the age of 20–30 also discloses nothing. What did you expect? In this initial seeding stage, stricken people were coming home to die, and the black population of Montana is 0.22 percent. There is not much you can do with such a "disclosure," even if you asked for it, since it represents no new information.

But in many cases the confidentiality issue lies far beyond reason, and beyond careful, thoughtful and sensitive analysis, because it has been mythologized into the sacred by those who wish to cling to their power in a human domain of life and death. And when power and myth take over, reason has little chance. What have been the reactions of various states and countries to the confidentiality issue and the release of useful geographic information? After a decade into the epidemic, North Dakota and Mississippi only feel comfortable reporting at the state level, while South Dakota for a long time took the bold step of publishing numbers east and west of the Missouri River, a physical feature that does not appear to be a barrier to HIV diffusion. Nebraska publishes figures by three economic regions, hardly the most relevant criterion for the geography of an epidemic, while Kentucky aggregates its 130 counties into 15 regions containing 5–17 counties each. Ohio reports on a 84 county basis, and Virginia by zip-codes. Finally, and as we saw above, Los Angeles has shown up the absurdity of most reporting by releasing numbers on a census tract basis, some of them only a few blocks in size. As we might expect, there is a general tendency for a state to report in smaller and smaller geographic units as the epidemic takes hold: South Dakota, for example, had a rate of 0.3 per 100,000, Ohio had 6.3, and Los Angeles 25.6. When it is finally too late to use information in the geographic domain it is reluctantly released.

It is no different in other countries, even those with some geographic education in the schools and universities. None of it seems to penetrate the corridors of power of the medical profession, which itself seems to have been elevated to the sacred. Such levitation constitutes an interesting and important intellectual history in its own right, one clearly dependent upon technological developments of all sorts that can prolong life and keep death at bay. It is not a story we can follow here – it deserves to be treated by fine scholars of intellectual and medical history – but it is important to note that the response to the AIDS pandemic is but a single example of a much larger trend and problem. In Sweden, for example, figures are reported by the three major cities of Stockholm, Göteborg, and Malmö, and then "Other" – in a country one thousand miles long and 100–300 miles wide – so that "Other" is essentially Sweden itself. Britain is little better, confining reported statistics to 16 regional health areas large enough to ensure that no one can do anything with the numbers. In a major government report, put out by the cream of the British epidemiological and statistical professions in 1988, there was not a single map, although a small table of the health areas was tucked away at the back. However, there was lots of conventional time modeling: one summary graph, based on England and Wales (no matter if Scotland had 20 percent of the cases), had 11 different curves spreading out from 1987 to predictions in 1992. These ranged from 2,000–12,000 AIDS cases, and one is reminded of the old cry of the fair busker "Yer pays yer money and an' yer takes yer choice." Unfortunately, by 1991, the actual cases reported did not lie particularly close to any of the curves, although by this time, five years later, nobody went back to check anyway. What would be the point?

Only Finland has monitored the HIV epidemic (as opposed to people with AIDS) on a geographic basis, using more than 1 million tests and 21 hospital areas as the units, and it is worth noting not only Finland's openness to education about sexually transmitted diseases, and its superb national healthcare system, but also the great trust the people have in the health authorities. Just as an example, in 1985 there was an outbreak of polio, generating a national health information campaign on TV and in the newspapers urging people to get an additional vaccination. Over 96 percent of the Finnish people complied on a totally voluntary basis, and the potential epidemic was stopped dead in its tracks by removing the susceptibles, the elements making up the backcloth carrying and transmitting the virus. It was the same story with the HIV: there were immediate information campaigns directed at homosexuals and drug users, with the full cooperation of these communities, and in the following years a similar campaign was directed at all travelers going abroad – over a million each year. By 1986, a huge campaign of public health saturated TV, newspapers, posters at airports and harbors, and a brochure was sent to every household in the country. The second phase was aimed at teenagers, in schools, on pop radio, in rock magazines, and so on, culminating with a booklet containing a condom for every 15–21-year-old. As a result, there is some very encouraging evidence that transmissions internal to the country have been drastically slowed down, most new cases being confined to travelers returning from abroad after vacations or business trips, the sources of the first infections. As of the middle of 1991, only 88 people had converted to AIDS, and while more will be reported through the end of the century, these remarkably open and trusted attempts to control the virus seem to have succeeded in large measure.

The issue of trust has appeared again and again in these areas of confidentiality, testing, treatment, reporting, and so on. There are good and obvious reasons for this. In societies with strong prurient interests in the sexual behavior of others, or in societies discriminating harshly against those infected, confidentiality protects the individual against the selfrighteous leer and the social rejection by others. In some places, and at some times, it is not an imaginary or inconsequential matter. We saw in Thailand (chapter 8) young couples being driven from their neighborhoods and committing suicide, and young women being forced to live in mean huts behind the relatively opulent houses their earnings as CSWs had provided for their parents. In the early stages of the epidemic, before many people are educated and aware, rejection may be particularly prevalent. It is the fear of confidentiality being broken that often makes voluntary testing in large-scale studies difficult to achieve. Once that trust has been broken it is often impossible to recreate it. And too many examples of broken confidentiality are available. One major longitudinal study reported that in one police precinct in a major American city a list of homosexual, HIV postive men was hanging on the wall – presumably, from the police point of view, so that precautions could be taken in the event of an accident or some other matter requiring police intervention. How exactly the police received the list no one will ever know, but leaks like these do occur, and send a shockwave of fear through such communities. Importantly, though, these rare cases of identification do not come about from geographic specification, although they make access to any relevant data even more difficult.

So apart from deliberate, and illegal identifications, the issue of confidentiality lies squarely in the geographic realm. Those (x,y) coordinates, unless aggregated to individual-protecting spatial units, can disclose identities, and most people would agree that this should be avoided at all costs. But aggregation to geographic units to protect individual people can be achieved depending on the geographic distribution and density of the population involved. But once aggregated into the smallest, but people-protecting units we can devise, what then? Is there any information in these maps? Can it be used to illuminate what is going on? And can it make a useful contribution to planning and intervention? Certainly none of the conventional epidemiological bureaucracies have shown they know what to do, although they have been sitting on such data sets for over a decade. So this is really the next question that we shall have to think about.

14

Education and planning: predicting the next maps

They don't have a plan – they wouldn't know what to do with it if you gave it to them.

A federal epidemiologist who must remain anonymous, December 8, 1989

The churches are grappling with AIDS – they're doing far more burials these days.

Panos Report, *6, 1989, p. viii*

What can we do? We have no cure, no vaccine, and at the moment these possibilities seem a long way off. The media may seize upon any glimmer of hope for a cure, of a chemical fix that will stop the ravages of the virus without poisoning the human body, but pharmaceutical researchers know they have a long way to go to hit those particular genetic strands of RNA without destroying everything else in sight. As for a vaccine, few realize how many years of carefully designed testing must precede the release and large-scale use of a vaccine with proven effects. We know now that the virus has the capacity to mutate relatively quickly, so that a vaccine produced to guard against one variety may be useless against another. Already trials for two vaccines are being prepared in Thailand, one for the "original" virus that formed the focus of early research, a second for a different form that appears to be more prevalent in the north. Meanwhile, in 1992, the WHO revised its estimates upward again, to 40 million HIV-infected people around the world by 2000, with the virus still raging unchecked through the populations of Africa, Asia and Latin America. Changes in protective behavior are agonizingly slow in coming, particularly for a constantly-renewed younger generation that sees little immediate evidence of the plague around it in the form of overtly ill and dying people. Most infected people have "moved on" a decade, and have little direct contact with those now reaching maturity. It is the nature of a slow plague to give no immediate signal that it is around.

Theodore Roosevelt once said, in his usual forthright and optimistic way that increasingly seems to belong to another era, "You do what you can with what you have where you are." So the question comes again: what can we do, and what do we have to work with? The answer is we can educate, and educate and educate; and we can do our best to help and comfort those afflicted now, and plan for those we know are going to arrive out of the future. At the moment that is all. And both education now and planning for the future make clear and unequivocal demands that any reasonable citizen can understand. First, we must produce educational materials that constitute the "cues to action" that make a difference, materials that reach out and grab young people. Second, we must plan to enlarge the limited medical facilities we have, facilities of all sorts – hospitals, specialized treatment centers, hospices and homes. Both tasks – education and humane caring – require that we estimate as carefully as we can *where* people will be. We need to predict the next maps.

Predicting the next maps is not an academic exercise, where "academic" implies the common meaning of irrelevant, useless, impractical, ivory tower, detached from the world, and pirouetting on the head of pin. It is not playing around in a computerized sandbox, producing numbers that no one can do anything with, but predicting where those numbers are likely to be in addition to when they might arrive. Predicting the next maps means incorporating them into educational materials, both in conventional printed form and animated television programs, so that young people can see where they are in relation to a raging epidemic. Predicting the next maps means being able to think about locating future medical facilities, and matching the size of those facilities to take into account existing capacities in order to augment them effectively and efficiently. "Effectively and efficiently" – some may blanch at bringing such criteria into the discussion, but the alternatives are "ineffectively" and "inefficiently." To enlarge and augment our capacity to care for people decently and humanely we need to use what we have in the most effective way. No country, no matter how wealthy, has unlimited resources. If you can make one penny do the work of two, so much the better. If you really want to serve people then you have to put those services as close to them as you can, and that means you must know where they are going to be. You need to predict the next maps.

Ways of predicting the next maps could easily take us into highly technical areas that are far from the province and intent of this book.

They require large and fast computers (a number of the maps in this book have employed big IBM and Cray machines), programmed to work through literally billions of those step-by-step computational procedures that we call algorithms. As a geographer, I cannot help pointing out that the very word "algorithm" comes from the name of the famous Arab geographer and mathematician Al-Khorizmi, whose step-by-step procedures became known to the Western world when his works were translated into Latin at the time of the Renaissance. There is a long tradition of mathematical analysis in geography, and we are on the frontier of modern developments when it comes to ways of predicting the next map.

But like many problems that are difficult to make mathematically well-defined so that a machine can compute the results, predicting the next map in a sequence is often intuitively easy for that lump of gray matter between our ears running on about 25 watts. The human brain is an extraordinary pattern integrator and analyzer, and if you do not believe me try this experiment. On the next page (no, do not look yet!), there is a predicted map of the AIDS epidemic in Ohio in 1993 (figure 14.1). It was computed in 1990 from the sequence we used before (chapter 6), and just to test yourself you may want to flip back to pages 68–9 and try to sketch on a blank map what you think it will look like five years after the sequence stopped. Whether you actually try drawing it or not, my guess is that you will be very close to the predicted map. Whenever we run our eyes over a map sequence showing the geographic diffusion of something, we seem to be able to pick up intuitively some sort of underlying spatial logic. The pattern on the map seems to be unfolding and developing in a consistent and predictable way, and we feel that unless something quite radically different and new appears, for example a cure or vaccine, that the process at work will probably continue for a while. All the ingenious complexities of computer algorithms are focused on the task of making this sort of natural geographic intuition well-defined and computable. But after all, transforming intuition into tested knowledge is what science is about.

Without going into all the complexities and computational details, what is actually involved? In many sciences observations are made over various stretches of time, and "time series analysis" is a well-developed and widely-shared approach in everything from astronomy through economics to zoology. Measurements in the form of numbers are plotted on pieces of graph paper or computer screens, and scientists look for some sort of order or structure in them, some sort of signal in the noise. In fact, that last phrase "signal in the noise" captures much of the scientific effort, because "noise" is the name we usually give to just random sorts of numbers over time. There seems to be no rhyme or reason to a noisy series, no trend up or down, no nice repetitive or periodic movements, like getting warm in the summer and cold in the winter, year after year after year. Things that happen quite regularly time after time after time are not random, and so allow us to predict.

Now predicting what is going to come at us out of the future gets pretty close to what the Greeks called *hubris* – it is an activity for the gods not mortals. Nevertheless, even if our prediction is only a bit better than tossing a penny, we may have a slight edge allowing us to prepare for what lies ahead. Presumably such soothsaying has considerable survival value for the human race as a whole, so it is hardly surprising that people have devoted a lot of thought and energy to forecasting – from diviner's bones, to stars in the sky, tarot cards, mathematical equations, and computer algorithms. Whatever approach is taken, we always try to find some sort of consistent pattern, and then assume it is going to develop in the same way a bit longer so we can extrapolate it into the future. Not too far, of course – that is for the gods – but perhaps a few years down the road before it disappears completely into the mist.

In analyzing a time series with all the methodological paraphernalia that has grown up over the past 200 years, we are really searching for information, the very opposite of randomness. Now geographers insist that in the same way that non-random information may be seen over time, so information may be discovered and analyzed over space. And in the same way that many scientists use rather sophisticated approaches to detect informative patterns in time series, so geographers have a number of tricks up their sleeves to find valuable information in spatial series – which is nothing more than a fancy scientific name for a simple and familiar map distribution. As you have just experienced in our little intuitive experiment, you can do quite a good job of picking out the information in a developing map sequence, and congratulations are in order because you have actually worked with something very difficult – a *spatio-temporal* series, numbers changing over space and time simultaneously. In one step you have left behind the traditional epidemiologists playing with their time series, and you are far ahead of conventional geographers who like analyzing single maps while forgetting time. Putting it in somewhat more pretentious "scientific" notation, you have taken the (t) of the time series analyst, and the (x,y) of the geographer, and have joined them together in an (x,y,t) spatio-temporal series.

Of course "spatio-temporal series," when said in a grave and serious voice, has a rather portentous, not to say pompous ring to it, but it is really nothing more than a pile of maps showing how the AIDS

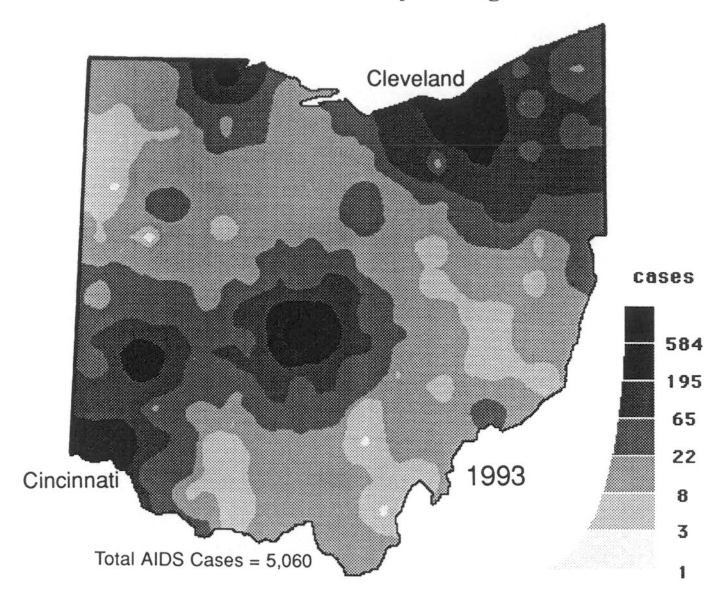

Figure 14.1 A predicted map of AIDS in Ohio by 1993. Notice, once again, that the interval between the contour lines and shadings is geometric.

epidemic is intensifying over space and time (figure 14.2). We can imaging a cube of numbers, each number representing the people with AIDS in a suitably small area (remember the matter of confidentiality!), whose center is (x,y), at a particular time (t). The trick is not to take just a thin sliver of one area along the vertical time axis of our cube, because that would just put us right back in conventional time series analysis. Similarly, we do not want to look at just a single slice along the spatial axes (x,y), because that would be just old-fashioned static geography again. What we really want to do is take the whole cube in one gulp and see if we can find the spatio-temporal structure in those numbers, the information that would let us forecast what the next maps of the epidemic are going to look like. Not too far, of course: geographers have developed some ingenious approaches to finding this sort of information, but with a few exceptions here and there they do not pretend to be gods.

How do they go about it? They start by taking the first map in the sequence, showing the first few numbers of people with AIDS and where they are. You may think that by taking the first single slice in the (x,y,t) cube they have thrown time (t) away, but we will see in a minute how time comes back. They then find other map distributions that seem to match the pattern of people with AIDS, perhaps something as simple as a population map. After all, the more people there are the

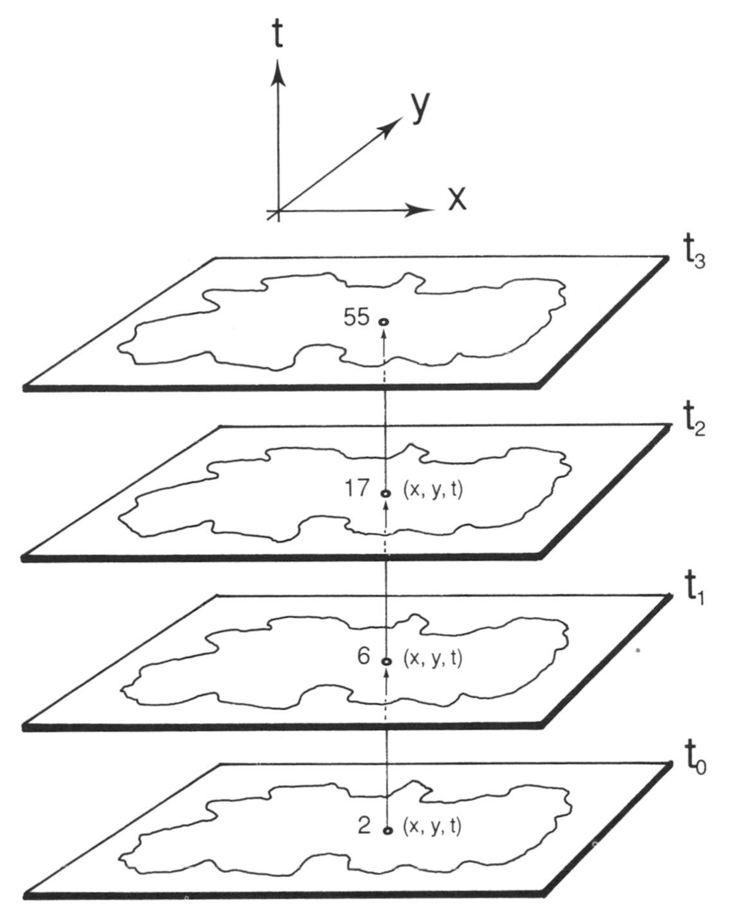

Figure 14.2 A spatio-temporal series in which the numbers of people with AIDS have geographic (x,y) and temporal (t) coordinates. A cube of such data, or a sequence of maps, contains valuable information to help geographers predict what the next few maps are likely to look like.

more likely we will find people with AIDS, because people are the only beings carrying the HIV. Later on we may want to examine more complex relationships, and see whether taking ethnic patterns into account helps (we have no evidence that it does), but it is better to start in simple ways, learning to walk before we run. And notice something: it is no use saying we might like to look at the match between AIDS and distributions of IV drug users or young male homosexuals. There is no way we could get these map patterns for obvious reasons, and we do not want to end up with our noses in the air claiming our theory has outrun the data. That is not a very helpful posture, and by adopting it you usually trip up and land flat on your scientific face.

As we might expect, simple population patterns match the developing distributions of people with AIDS quite well. In densely populated urban areas where there are lots of people we find many with AIDS; and in sparsely settled rural areas we find only a few. So far so good, but you do not need any fancy computer algorithms to say this sort of thing. What we actually have is a very general and overall relationship between AIDS and population expressed as a little function. Without getting too technical it might be a cubic equation describing a line that can both rise and fall. In the early stages of an epidemic we may have an accelerating (bending upwards) relationship, while later the large cities may experience a slowing down (bending over) sort of relationship. At the very least we want to allow for such possibilities. A geographer calls a function like this a *filter*, because in some sense it accepts some values and not others. It is actually a *spatial adaptive filter*, a rather fancy and complicated name at first sight, but really quite accurate as we shall see in a minute.

I want you to image that our functional filter is like a small geographic bloodhound that we can unleash, allowing it to wander all over our map sniffing around for spatial structure or information. If our geographic bloodhound finds anything, it adapts its own numerical form to these local conditions, giving a much better fit and prediction for the immediate area than the overall filter that we started with. If you happen to have had some experience with systems analysis, you will recognize that this sort of thing always requires some sort of negative feedback going on at each step to help the adaptive process. It can lead to all sorts of technical complications, most of which can, for all practical purposes, be overcome quite easily if we have large and fast computers. Fortunately we do. As for our purposes, they are very practical indeed.

So what we end up with, after our geographic bloodhound has wandered around our first map in the sequence, is a functional filter that has used the spatial information in the map pattern to adapt to local conditions – hence *spatial adaptive filter*. It can now compute predicted numbers of people with AIDS that are very close to the values we already know from the history of the epidemic. Of course, at first sight this may seem to be sheer madness. After all, why go to all that trouble making your little bloodhound filter adapt to local geographic conditions to predict values you already know? What is the sense of trying to "predict" when you have the numbers there staring you in the face? But there is actually method in such apparent madness. For each of the areas on the map we keep a record of the adapted values, the coefficients of our adaptive filter producing those close and locally meaningful "predictions." We now take the next map in the sequence and do the same thing, and then the next, and the next, working our way, geographic slice by geographic slice, up our (x,y,t) cube along the vertical time axis. For each time slice we record for each area the coefficients in our filter that have allowed it to adapt to the local conditions using all the emerging and strengthening geographic information in the map pattern. It is really not that complicated, and only what you did intuitively as you ran your eye along that map sequence, feeling that emerging pattern getting stronger and stronger, and so allowing you to make a good guess at what the next one, or two, or three maps were going to look like.

So the next step is simple: we take each area in turn, and see how our geographic bloodhound has adapted itself over time, how its coefficients have altered by adapting each time to the changing geographic pattern of the epidemic. If the epidemic had spread in a totally random fashion, something we could imitate by pulling random numbers out of a hat and plotting them on the map, then our poor bloodhound would have had a terrible time adapting because there would have been no coherent spatial information available. In this case, when we take a single area, and look at the way the filter's coefficients change over time, they just bounce up and down in a chaotic way, reflecting the no-rhyme-or-reason progression of the random map sequence. Fortunately, in our series, those adapting coefficients do not bounce up and down. They actually track very smoothly over time, changing in highly regular and therefore predictable ways. All we have to do is estimate what they will be for several time steps into the future, and then use them to predict the intensity of the epidemic at that time *and at that place*! We can predict the next maps.

There are other ways to predict maps – transformational, expansion and neural methods – but this is not a technical book, and it is much more important to get the sense and feel of one of these geographic approaches than all the technical details. The important thing is that we can get a pretty good idea of what those next map distributions are going to look like, and while some of these approaches are fairly new, others have been available for decades.

By ignoring, not to say suppressing, the geography of the AIDS pandemic, country after country has thrown away a fundamental perspective with totally practical implications for programs of educational intervention and the proper planning of extended healthcare. We have already noted many times the way in which maps, in simple sequences or in animated form for television, can literally "bring home" to young people the deadly dynamics of this pandemic, making them realize it is not something "out there" and remote, but literally all around them. We do not have such a plethora of those cues to action that we can afford to dismiss any possibility to enhance the educational message.

As for health planning, this could lead us into completely different areas of geographic knowledge, areas well developed over the past thirty years as computers gave us an increased ability to solve difficult problems of spatial optimization and allocation. Said plainly, if you want to increase the capacity of any healthcare delivery system, you *must* take into account the existing geographic pattern of hospitals, outpatient facilities, hospices, and so on. There may be some areas with an excess capacity, others feeling great stress and overcrowding. We have to balance out needs and requirements, as well as increase the overall capacity of the system, and that "balance" is always a *geographic* balance. Geographers have developed many computer algorithms to find optimal patterns of allocating new facilities, but they all require as a basic input of information *where* you expect the people are going to be who will need them. Whatever the geographic scale, you need to get some handle on what those maps in the future are going to look like.

Quite tragically, those increases in our ability to deliver decent healthcare are not needed simply to treat people with AIDS. What is increasingly clear is that as many individual immune systems go down, so other diseases, long since thought to be under control, begin to come up again. We glimpsed this ability of another disease to ride on the coattails of the HIV with the rise of tuberculosis in children in the poorer areas of New York City. As the immune systems of HIV infected people go down, so the TB bacillus, normally held at bay, begins to thrive again. Infected adults carry it to overcrowded and unventilated buildings to children, for it is often transmitted in the aerosol particles of a cough or sneeze – or simply a breath. As the immune systems of many people decline, so the immunity of the population as a whole decreases. Veterinarians talk of "herd immunity," and we must think about this too in the context of human populations.

15

Herd immunity: riding the coat-tails of the HIV

> Just as a human body with some defect in the cellular units that confer protection against disease suffers by becoming diseased, so, too, when a society's individual components have immune defects, the entire society can be seen to suffer a morbid consequence.
>
> *William O'Connor, MD, Vacaville, California*

Most people instinctively shy away from the term "herd immunity," particularly when it is applied to people. Referring to a population of human beings as a "herd" leaves us uncomfortable, for the word has that disturbing reifying ring to it we met before, implying that we are prepared to think about people as animals. But like many terms in human and veterinary medicine it has a precise technical meaning in its own context, and in that context it carries none of the perjorative, not to say unethical overtones of meaning. It means simply that if we take a large number of people we can think of them possessing collectively some general level of immunity to certain diseases. In a very strict and biologically defined sense we are part of the animal kingdom, and in exactly the same way that certain populations (herds) of animals have a collective resistance, so do collections or populations of people. We met the former example (chapter 2) when we saw that some simian immunodeficiency viruses were perfectly well dealt with by the immune systems of green monkeys and chimpanzees in Africa, even though they were deadly to macaque monkeys in Japan. As far as humans go, we have both archival and archaeological records of the terrible devastation caused by diseases like measles that were brought by Europeans to the New World, only to move like a forest fire through populations whose people had virtually no immunological resistance to diseases they had never experienced before.

So while acknowledging that words may carry different connotations depending on the particular context, let us see what "herd immunity" means in the scientific sense we are talking about here. After all, every child immunization campaign around the world has as its goal the raising of the collective or "herd" immunity, and we saw in Finland (chapter 13) how 96 percent of the people responded voluntarily to the vaccination program for polio, so stopping a potential epidemic in its tracks by raising the immune level of all. After the vaccine had raised the herd immunity of the Finnish people to a high level, the polio virus had no backcloth of susceptibles on which it could exist. And this is precisely what herd immunity means: it is the ratio of those people who are resistant to a disease to those whose immune systems cannot cope with an infection. If a society is going to maintain a high level of public health, it is obviously of crucial importance that it thinks about its collective responsibility. That is why we have Public Health Departments, Federal Health Centers, health ministries, and other governmental institutions charged with the responsibility of monitoring and doing something about these things that have the potential to affect all members of a society.

Now it is obvious to the point of tautology that the herd immunity of a people ultimately depends on the ability of the immune systems of individuals to handle infections. Most of the time the immune systems of healthy people – augmented by polio, influenza, measles, mumps, scarlet fever, cholera, and many other immune-boosting shots – deal with a large number of infectious diseases. And even if some of these do appear from time to time, most of them can be contained and cured by antibiotics and modern medical practices. For many people around the world today the late nineties of the twentieth century are very different from the times of their great grandparents. We tend to take this enormously enhanced ability to raise the herd immunity for granted. Perhaps too much for granted. When an immunosuppressive virus like HIV moves slowly, and at the moment with seeming inexorability through a population, many individual immune systems collapse, so allowing diseases otherwise normally kept in check to become prominent once again. As we have seen repeatedly, AIDS is precisely that condition when the immune system cannot cope any longer, and conversion is always to the so-called opportunistic infections. What is so dangerous about the overall decline in herd immunity is that more and more people become carriers of these diseases, and so possible sources of transmission to healthy people.

This is nothing new. In the old phrase "Typhoid Mary" we acknowledge that people can be carriers of diseases while not showing any symptoms themselves, and for the sake of others we feel that they should be identified and disconnected until their infection is cured. Every year common colds and influenza viruses sweep through our societies, and we have often said to someone "Why are earth didn't you stay home today?", implying that they should have taken themselves out of circulation for a while to prevent them spreading their current malady to those around them. People preparing and handling food eaten by others are especially potent sources of potential infection, and every restaurant has signs reminding its employees to wash their hands and observe high standards of basic hygiene. In all these cases, you have to break the connections over which the disease can be transmitted; or, in our structural language, you have to fragment the backcloth (chapter 4).

One of the things we have observed over the past 15 years, and are observing with increasing frequency today, is a rise in many other infections parallel to the rise of HIV. Recall that with a median time of ten years between initial HIV infection and conversion to AIDS, the virus was starting to move through particularly vulnerable populations (homosexuals, IV drug users, blood transfusees, hemophiliacs, etc.) in the seventies. HIV is a slow but distressingly sure plague, and although the first few signals were hard to detect, many are beginning to wonder now if the slow decline in the collective immunity of a people was not beginning to show even as early as the mid-seventies. It is a difficult question to answer with absolute certainty, because things other than a immunosuppressive virus can lower herd immunity. For example, in the Third World many diseases are chronically endemic, and they lower the resistance of people, especially children, to a host of other infections. Malnutrition, lack of proper healthcare, substandard and overcrowded housing only make things worse, and we do not have to go to the Third World to see these conditions. Nearly all of America's inner cities have health statistics (for example, infant mortality) that are as bad as many poverty-stricken Third World countries. If you essentially shred the social safety net of a people over 12 years while HIV is moving through you cannot expect things to jog along as before. Your claim to be a caring and civilized society may also be placed in doubt.

What evidence is there of other diseases increasing parallel to the spread of HIV, and what diseases are involved or strongly suspected? Interestingly the evidence is not just over time – a marked increase in other infections during the eighties and early nineties – but also over space. But we have to be a little careful here. Neither simple increase over time, nor equally simple spatial juxtaposition, in and of themselves constitute strong scientific evidence of cause and effect that people like to have. Just because two things, say the sales of umbrellas in Tokyo and scotch whiskey in Stockholm, happen to rise and fall together over a period of time does not mean that one causes the other, that the haze of alcohol in Sweden causes rainfall in Japan. The ups and downs maybe coincidental, or they may reflect some other thing entirely. Similarly, just because two things occur at the same place, in the same geographic area, does not prove they are causally connected. On the other hand, and despite these warning qualifications, we do not have to adopt a totally skeptical or moronic pose and throw our commonsense out of the window. And if we are scientifically trained, we know from many research investigations that it is precisely things moving together over time, or appearing together in space, that often catch our attention and underlie those initial hunches or hypotheses that lead to the most fruitful scientific insights. That is why geographers are forever putting things on maps, and making those map sequences that show changes over time and space.

Yes, yes, you say to me, but so what? What's the point? Well, because it is precisely in this arena of parallel temporal movement and spatial juxtaposition that we are going to see the epidemiology of the AIDS epidemic as an example of science as a socially contested enterprise. Two groups of people are going to look at the same epidemic, the same figures, the same "text," and they are going to interpret them in quite different ways, placing quite different meanings on them, particularly in the light of the possible logical consequences that might follow. It will become quite clear where I stand. I do not believe in certainty any more – that is for the gods not mortals – and in the midst of a raging pandemic I am not prepared to wait around. Plain old "strong possibility" is good enough for me.

Selfish – maybe. But I am not the sort of person who is prepared to stand around just to see how many people die from radioactive fallout amplifying up foodchains, nor do I want to while away the hours to satisfy a casual curiosity about how many skin cancers or eye cataracts develop as the ozone hole gets bigger. In the same way, I am not prepared to take the supposedly detached view of the ideal scientist and wait around twiddling my thumbs until a series of huge epidemiological studies over many years demonstrate to all but the most obdurate that, yes by golly, it really does look as though the lowering of millions of individual immune systems allows other infections normally kept at bay to increase and result in millions of deaths.

Millions? Exaggerating? Is this guy "off the wall"? Well, the WHO – hardly a wild-eyed radical body of international bureaucrats – has, as we have noted, estimated 40 million HIV infections worldwide by 2000 – a date not so far away. In the United State alone we have between one and two million people infected, and no end even remotely in sight at the moment. And the CDC, and health committees in Congress, not normally distinguished for their ability to move with dispatch, now estimate 10 to 15 million Americans are infected by the TB bacillus. "Although," said one of those anonymous spokesman, "fewer than 10 percent will develop an active case in their lifetimes." Oh well, what's a million or two cases here or there? A pity the drug-resistant strains are increasing though.

So with these very careful qualifications in mind, what other diseases might be involved and coming up as the herd immunity of the American people goes down? I am going to focus on the possibilities in the United States, since there are literally hundreds of professional medical articles to draw upon and piece together, and the country is already moving towards 400,000 people with AIDS (under the new definitions), and . . . well, who really knows how many HIV infected people there are between those 1–2 million estimates? In Africa, Asia and Latin America, we do not have anything like the same set of medical studies to draw upon, but I see no reason to suppose that the story will be significantly different.

One disease that has appeared has a name that you have to say slowly the first couple of times you meet it. *Cryptosporidium* is a protozoan, a tiny single-celled animal that causes chronic and almost unstoppable diarrhea in people with AIDS, and was one of the causes of a condition labeled the "gay bowel syndrome." It appears in 15–20 percent of the cases with this condition. Before 1976, a time when the HIV was already on the move, it never seems to have been observed by doctors, and was never reported in the medical literature. But in the five years 1976–81, five of the first seven cases of *cryptosporidiosis* ever recorded in medical history were in people whose immune systems had already been severely lowered by HIV. Then, sporadically throughout the eighties, it began to be reported in childcare day centers, particularly in California. It is a highly infectious disease which can spread from person to person directly, through the air as dry spores, and through water. During the late 1980s it infected more than 12,000 people in western Georgia (recall the band of high rates and accumulating peaks of people with AIDS, chapter 9, figures 9.4–8), as a result of a contaminated water supply, and it is particularly dangerous in children who can die from the almost cholera-like effects of dehydration. Once again, and you may reach a point quite soon where you are weary of this repetitive caution, the rise of *cryptosporidiosis* could be a coincidence, could be simply a result of the breakdown in public health and hygiene infrastructure, and so on. But it seems a perfectly reasonable hypothesis that it appeared first, and spread rapidly in the homosexual communities who took the first brunt of the pandemic. These communities constituted a series of local populations with compromised immune systems allowing the protozoan to accumulate to the point that it constituted a significant pathogen for the first time in American medical history.

A second disease rising rapidly in the eighties was hepatitis A, increasing 58 percent between 1983 and 1989. Back in the sixties and seventies, hepatitis A was a disease associated with poor people living in areas of crowded housing, with poor sanitation facilities, and practicing levels of personal hygiene that enhanced person to person transmission. It was the sort of thing most people realized you could pick up if you were not careful in Third World countries where the food and water might be contaminated. But we now have a minor epidemic of hepatitis A, the most prolonged outbreak this country has ever seen, and again and again the source cases seem to be people with severely compromised immune systems. In Florida, for example, an outbreak of 103 people with hepatitis A was traced back to food handlers working in the pantry of a restaurant, two of whom were frequently-practicing male homosexuals, one of whom came down later with hepatitis B. Texas and Alaska also experienced similar and sudden outbreaks. The former in a salad bar in which the sources were male homosexuals preparing and handling food, the same sources as in the Alaskan case. In California, almost half of the hepatitis A outbreaks traced back to food involved preparation and handling by homosexual men. There is, of course, no compulsory testing, so the direct immunosuppressive link cannot be directly established. We shall have to examine such reluctance to test in the face of the strong possibility of one epidemic riding on the coattails of another.

A third disease increasingly associated with AIDS is a series of bacterial infections called the *Mycobacterium avium* complex, or MAC for short. For a long time it was thought that the rare cases of MAC were due solely to transmissions from the surrounding environment. The bacteria can remain alive for long periods of time, and they can be spread through the air or by contaminated water supplies. Most people with healthy immune systems easily cope with any chance contact, although contrary to what many non-specialist doctors think, perfectly healthy people do come down with MAC from time to time. But until the eighties most doctors could go through a lifetime of general practice and never see a case. Today, 53 percent of the autopsies performed on people who have died from AIDS have MAC as one of the several so-called opportunistic infections. Moreover, when it is identified in people today through careful examinations of their sputums, the MAC strains are usually different from those in the immediate environment, suggesting strongly person to person, rather than rare environmental transmissions. The bacteria are extremely difficult to treat, and often result in death despite very broad spectrums of antibiotics being used. Healthcare workers appear particularly vulnerable: contaminated instruments were the cause of 17 transmissions, and the bacteria have contaminated water supplies in hospitals treating a high proportion of AIDS patients.

Once again, the suppression of the immune systems of large numbers of people by HIV – remember, 53 percent of 368,000 is nearly 200,000 people, most of them highly concentrated geographically – is not something that can be airily dismissed. This is a particularly complex part of our complex world, and single causes and effects are not easy to see or disentangle. When several causes work together, changing their relative causative strength over time and over space, the nice clean laboratory experiments of the physicist and chemist do not help us any more, and the clear and always theoretical experimental designs of the textbooks only demonstrate how far human theory may be from human practice. Many statisticians either throw up their hands in despair at ever teasing out these webs of multiple causality, or demand huge and long-term epidemiological studies that move quickly beyond the bounds of cost and practicality. But these difficulties should not close our *thinking* to these possibilities of complexly interacting causal factors, one of which may be the lowering of herd immunity, especially in particular regions and at particular places that the geographer doggedly insists are important to consider. Not the least because these places too often become the regional epicenters from which hierarchical and spatially contagious diffusion occur. We shall see this in the catastrophic rise over time, and at concentrated juxtapositions over space, of tuberculosis.

At the risk of annoying you one more time with cautions and qualifications, we are entering here an area of enormous complexity, one where multiple causes become so convoluted that it becomes – I will stick my neck out again – *impossible* to separate all of them out cleanly, assigning such-and-such percentage of effect to this, and such-and-such a percentage of effect to that. When it comes to the individual case, the etiology – the cause and origin – of tuberculosis is one of the most complex in medicine, and a proper understanding requires that we abandon that old Cartesian dualism of mind (whatever *that* is!) and body. People are people, whole beings, not two bits of seventeenth-century speculation stuck together.

In the last analysis, tuberculosis, TB, is "caused" by chronic infection by the bacillus *mycobacterium tuberculosis*. It is not rare – about five percent of the American people carry it – but in most cases this *primary* TB is kept well under control and in a dormant, non-infectious state by a healthy immune system. In the nineteenth and early twentieth century it was the scourge called "consumption," from which poor people died relatively quickly, while genteel Victorian ladies faded away coughing their lungs out, or went to expensive clinics at high altitudes in Switzerland. After the Second World War, and well into the seventies, we thought we had it licked – at least in the First World. Rising standards of living and healthcare, and the fact that it responded well to the new antibiotics, resulted in decreasing rates year after year. In the United States, for the thirty years after 1953, when we really had a uniform reporting system for the first time and could monitor it properly, TB declined by an average of 6 percent each year. Even during the sixties and seventies, both the numbers of people and the rate went down and down, despite the fact that many refugees from south-east Asia were infected.

Then, in 1975 in New York, the cases started to go up, and so did the rate. Possibly it was only a random blip – after all, the numbers and rates continued to go down in 1977 and 1978. But the decline did not last: from 1979 until the present, the TB rate in New York, and in other metropolitan areas across the country, has risen steadily. Between 1984 and 1986 in New York alone it rose 36 percent, and kept on climbing to 38 percent between 1990–91. We have already seen the localized but catastrophic rises in TB in young children (chapter 10). Quite tragically none of this came as a surprise. In the late seventies and early eighties people in public health, chairpersons of innumerable task forces, social workers and human ecologists had recognized the threat and were calling for more funds – funds canceled by the Reagan administration, restored by the House Health Committee, but never funded by Congress. TB control in New York went from $40 million in 1969 to $2 million a decade later. CDC was hog-tied by Congress, whose members let TB funds for particularly hard pressed areas decline as low as $1 million between 1982 and 1989. It was like turning a lawn sprinkler on a forest fire.

The raging TB epidemic coincided almost exactly with the time the social safety net of the American people was pulled out from under them by their own politicians. After all, healthcare was essentially an individual matter, and the private sector, with its experienced and highly profitable insurance companies, was the best forum in which the fine hand of the market could operate. Mind you, it might be necessary to do a bit of token public health here and there, but not too much. People must learn to be self-reliant. Only when the advancing rates reached levels even Congressmen could understand, and drug resistant strains appeared, did federal funds become slowly available. After all, even Congressmen and middle-class people could now die of drug resistant TB.

Was the TB epidemic *caused* by HIV lowering the herd immunity of the people? This, by itself, is too simple. We know that as social conditions decline, as overcrowded and badly ventilated housing becomes the lot of more and more poor people, that the probability of transmission rises steeply. We also know from many medical studies focusing on TB, that immune systems become less and less functional under psychic stress. With slum conditions, burnt-out areas, population displacements, and the fragmentation of those supportive family and neighborhood human networks, the decline in localized herd immunity from psychic stress will result in TB carriers going from the dormant to the secondary and active stage. This is why I said we have to abandone that old and useless dichotomy between mind and body. But even this too is too simple, too facile. If, at the same time that you have chronically deteriorating social conditions, you *also* have the increasingly rapid spread of an immunosuppressive virus, then you can only compound and amplify the conditions for an explosive rise in TB. Immune systems already under psychic stress start a freefall as the HIV begins its work, so you augment the pool of carriers shifting from the primary to the infectious secondary stage, spreading the bacillus even more widely. After all, people travel – around the city, from city to suburbs, from city to city. TB, like HIV, like *any* infectious agent carried by people to people, is going to spread by jumping around the map controlled by hierarchical diffusion, and by oozing out of the epicenters by spatially contagious diffusion. Even the US Public Health Service recognized that complex cause and effect relationship as early as 1987, relating part of the rise of TB to HIV; and in 1992 CDC shifted $15 million from HIV-prevention programs to TB control. "Given the very close link between HIV and TB," said one of those anonymous spokesmen, "we're not raiding [the AIDS program and budget] but achieving a synergy."

But HIV and TB had already achieved a "synergy," a mutually enforcing cause producing a complex effect, 15 years before. The problem was how to tackle these devastating effects in the face of equally devastating public complacency that translated politically into an indifference to the growing squalor of America's cities, and an unwillingness to commit funds to deal with a truly public, as opposed to a private and individual, health problem. As the social safety net was jerked out from under, and TB funds were continuously cut, the increasing number of active TB carriers became more and more difficult to treat. The course of drug therapy is a relatively long one, a 6–12 month regime of regular and carefully spaced treatments. In poor areas it requires almost person-to-person monitoring to make sure the full course is given. But even by 1982, one doctor in New York noted that there were no longer the resources to monitor many active carriers, some of whom just wandered away half-treated to provide almost ideal conditions for the emergence of drug resistant strains.

Other diseases have also been on the rise during the eighties – salmonella, syphilis, measles, influenza, and so on – often appearing at the high peaks of HIV infection on the map. AIDS patients are about 20 times more likely to be salmonella carriers than non-infected people, and are very difficult to treat. Syphilis, in steady decline since the introduction of antibiotics, reversed its trend in 1987 with a 25 percent rise, with more than half the cases in New York, California and Florida, once again the soaring peaks of the HIV and AIDS epidemic. Measles was also in decline until 1983, when the trend reversed, and in 1988 nearly 70 percent of the cases occurred among junior high and high school students, 12–19 year olds who had been vaccinated as young children.

But now comes part of the contestation: in all these cases and others, alternative explanations can be, and have been, put forward, other than the possibility of a general lowering of the population's collective immune response. At one level this is perfectly appropriate: there are famous scientific papers on "multiple hypotheses" and why it is important to keep thinking open to alternative possibilities and explanations. Salmonella somehow gets into battery laid eggs; syphilis increases because of increased sexual promiscuity in the age of the pill; measles rises because we are shamefully behind even some Third World countries in our child immunization program and requirements; and everyone knows influenza fluctuates up and down anyway, although some doctors now recommend that all those in immediate contact with HIV infected people receive annual immunization shots. In a complex process of multiple causes, the mere *possibility* that the herd immunity of people might be one, but perhaps important, causal thread in the complex epidemiological tapestry is dismissed, denigrated, and ridiculed. Why? Why the contested nature of this most plausible hypothesis?

Because if you are prepared to take the possibility seriously, you raise questions that have become an anathema, virtually unthinkable, in today's highly politicized AIDS arena. Questions, essentially forbidden questions, arise again as thinking moves once more to the age-old and always contested question of the rights of the individual and the rights of the larger society. "Compulsory testing" and "quarantine" have become what the distinguished literary critic George Steiner called "night words" – words and phrases that are no longer allowed to enter discussions. Even those who approach such questions obliquely, using euphemisms and circumlocutions, generate an immediate and highly heated response, and they often become the target of abusive accusations to the effect that they are no better than the guards of a Nazi concentration camp. One doctor in California, a specialist in preventative medicine in daily medical practice, was excoriated by his colleagues at a famous medical school for even daring to think such thoughts, let alone publish them in a highly reputable, peer-reviewed medical journal devoted to preventative medicine – medicine, in other words, dedicated to stopping things before they start, usually at a cost orders of magnitude below the price of treating a fullblown epidemic. "Ten years from now," he said, "I shall probably be told I was right – but went about it the wrong way."

How has even the possibility of raising such a question become so highly charged emotionally that it distorts medical practice, scientific study, and the political process itself? It seems to me that there are two elements here. First, recall the difference in human response to quick plagues and slow plagues, to serious, perhaps mortal diseases with fast versus long incubation times (chapter 4). Cholera, typhoid, yellow fever, and in past days smallpox, all required an infected person to be disconnected from the larger social backcloth to protect others. No one questioned that it was the right and responsible thing to do. To do otherwise, to ignore deliberately quarantine as a response, would mean many others would suffer. Such irresponsibility would be like the HIV infected man who said "I'm gonna take as many with me as I can." No society concerned with its very existence can allow such sociopathic behavior in its midst. But with a slow plague the response is different. We never see the immediate effects of transmission, and we know transmissions of HIV are rare in casual, everyday contact. Most of them have to be much more direct – blood, needles, unprotected sex. And for all the ills of our society, it still has a great capacity to display compassion, and demonstrate a sense of caring and decency arising from a strong ethical sense of what is right and wrong. Much of it is there in the Constitution, whose constant and on-going interpretation is yet another arena of contestation as an eighteenth-century ground tries to bear the weight of a twentieth-century world.

In the context of the AIDS epidemic, much of the ground contested has been occupied by the homosexual community that took the devastating effects of the first HIV wave. And here, no matter what our own sexual proclivities may be, we must try to *understand* before we think beyond the immediate situation to future possibilities. For all its often strident bravado, its solidarity marches, its acting up, it is a community psychically shredded and in spiritual agony. In the seventies it was a community just emerging with its own identity from a highly conservative and rigidly conformist society. In many respects, particularly in rural areas and in certain cultural groups, sexual attitudes are still that way. Even among young adults, 20 years later, degrees of tolerance towards sexual diversity differ widely. In that most intimate area of human behavior, the sense of right and wrong, what is acceptable or unacceptable, is likely to be felt very deeply. I know – I see the wide spread of opinion everyday in a large public university.

For the homosexual community the seventies were a time of emergence, of small but growing gains in tolerance, particularly from the better-educated people of the larger society. Old laws were dropped from the books, and a general opening up of heterosexual behavior through the contraceptive pill seemed to herald a greater tolerance of other forms. And then in the eighties, just as closets were being opened, the terrible effects of those HIV transmissions back in the liberated seventies struck. To lose many of your friends suddenly in an accident is bad enough; to lose them slowly, to see them slipping away, often in agony, tears your own soul apart. For many, what started as love became death. Bewildered, angry, frightened, asking "Why me?", "Why him?", the community had only one thing to hang onto – itself. One face of the suffering was the deep compassion, care and love that the community found for its own and others; another was the anger that translated into "them versus us," into a what-have-I-got-to-lose militancy that became political action.

Much of the political action was directed at any move on the part of the larger society to discriminate in any way against the HIV infected or those with AIDS. And I think in the view of most decent people, properly so. As the eighties progressed, and we learnt more and more about channels of transmission, it became clear that the virus was not easily transmitted. Quarantine was out of the question, unthinkable and impractical, and in light of the minute probabilities of transmission outside of the identified channels many thought compulsory testing served little or no purpose. Moreover, it ran up against some of the deepest feelings about individual rights, and the general distaste for authoritarian gestures in a society that prided itself on individual respect. Compulsory testing also brought to the fore the question of individual privacy, and how breaking confidentiality could be grievously misused by employers, insurance companies, and so on. Anti-discrimination laws were passed at the state and federal levels, and confidentiality was enforced so strictly that it became virtually impossible for even medical people to know what was really going on. If you test blood from volunteers, and strip away every vestige of identification, you cannot even tell someone they are infected, you cannot counsel them, and you cannot help them prevent further transmission. We still do not know what the general level of HIV infection is because all attempts to carry out a large and properly designed epidemiological study have failed at that first political hurdle.

At one level I think all this is understandable, but it may be literally vital for many in the future that we understand why we are in the situation we find today. At least that is the first step in getting ourselves out of it. A mortal pandemic is a dynamic thing, it does not stand still, and neither can our thinking. If it were just (?) a case of HIV and the knowledge we have of it today, we might feel that the issue of compulsory testing should remain one of those night words. Unfortunately, it is not "just a case of HIV" any more in light of the appearance and rise of other, much more easily transmitted diseases, as hundreds of thousands of immune systems decline. These shift the collective immunological response of the populations down, often way down in populations highly concentrated geographically forming the "point sources," the regional epicenters, for further diffusion. Given that we will be approaching half a million people with AIDS by the time this book is published, I think the issue of compulsory testing has to be raised again under very carefully delimited and circumscribed conditions that can continue to guard the strict confidentiality of the individual. This is contested ground, as it has to be in a genuinely democratic society, but even in the most liberal, even anarchic society, the rights of the individual always have to be thought through in relation to the rights of the larger society.

And testing, of course, is going on, sometimes contested, but in many cases not contested at all because the conditions have been carefully delimited and circumscribed. Blood must be tested today if it, or some fraction of it, is going to be transfused to others. Suppose, for example, Dad needs an operation and has an uncommon blood group. Close family members will volunteer to give blood, which will be tested. Suppose you are a young homosexual, or have used intravenous drugs at some point, but no one in the family knows? Are you infected? Up to this point you do not know, and may not want to find out, but now you find yourself under great moral pressure to give blood to someone you love. If you refuse, members of the family will ask why. If you give, your blood must be tested before it can be used in a transfusion. To handle this problem, most hospitals require family and friends to fill out a form under conditions of total privacy and guaranteed confidentiality so that those who have even the slightest suspicion of infection can indicate this. Blood can be donated so that other members of the family do not suspect anything, but the blood is promptly thrown away – untested, unless the donor specifically requests it.

This particular case seems clear enough. But what about testing, for the protection of the larger society, those diagnosed with TB? Given the temporal and spatial juxtaposition of HIV and TB today, is it not the responsible thing to do? Given the climate of opinion, backed by political demonstration and action, many in the medical profession are still reluctant to take such a stand. But what about certain occupations where people whose immune systems are going down are in direct and often intimate contact with others? For example, people handling food, in testing labs, in nursing, childcare, and so on. Many companies are testing for drugs in the work place, why not HIV? Is not an ounce of prevention worth . . . well, not a pound of cure, for there is none, but at least other human lives? When there are outbreaks of infectious diseases traced back to food handling, should there not be HIV testing of those identified as carriers? Is this not the right and proper thing to do?

And to go a step further, what do you do in the case of TB when HIV compromised people often do not show up on standard tests? Tests in Zaïre indicate that nearly one third of the HIV infected people tested for TB had no response to the standard skin tests, even though they had the active form. Even chest X-rays may be ineffective as alternative diagnostic tools because the classical cavitations in the lungs do not show up and special tests must be employed. Finally, in a population with many individual immune systems going down, the chance of new mutant forms occurring rises steeply. In a people with a high collective resistance, most of these mutant forms would be dealt with immediately by healthy immune response, literally "strangled at birth," rather than being allowed to accumulate into more resistant forms to be transmitted to others.

If it were not so tragic, we could label such a situation at least as ironic. As a society, people are prepared to think about raising the level of collective immunity, and take collective, that is social, action to lift it up. Every child immunization program has this as its laudable aim, every call to come forward for booster shots when there is a local outbreak has this as its goal. And everyone says "Of course, this is the right, rational and sensible thing to do." So why is there such reluctance to *think* the reverse, to think through the consequences of a move in the collective immune response downwards instead of upwards? Why has this become such contested ground producing such highly charged and emotional responses? Where does my right impinge on yours, and yours on mine? These are questions we must be willing to ask and think about at the very least.

16

Epilogue: old plagues for new

There is a mistaken public perception that AIDS is a worldwide disease that is spreading. The AIDS problem has been distorted and exaggerated. AIDS was most commonly transmitted by anal intercourse and intravenous drug use.

Dr Albert Sabin, developer of the oral polio vaccine, quoted in The Bangkok Post, *September 21, 1991*

The signs of indifference are already apparent. It was possible to sit through six intensive days of AIDS research at the seventh international conference on the disease at Florence last week and feel a growing sense of unreality.

Christopher Mihill, "A deadly division," *Guardian Weekly*, July 7, 1991

And The Band Played On

Randy Shilts's title to his book on the AIDS epidemic

Perhaps you have already sensed the contradiction in any attempt to write an epilogue to match the prologue we started with. An epilogue implies some sense of closure, a looking back and reflecting upon something that is past and behind us. How can we have any sense of something closed and finished when we are still only at the beginning? But I know of no other satisfactory way to end this book, except perhaps to finish in mid-sentence. "Summary" and "conclusion" seem equally inadequate, and again imply a looking back on what has been, when all our eyes and thoughts should be on and open to the future.

Not many eyes, and perhaps still fewer thoughts were open back in 1982 when the HIV-1 was first identified in the 1,485 people then diagnosed with AIDS in the United States. Who then would have forecast nearly 400,000 cases (under new definitions) a decade later, or think that the WHO would be guesstimating 40 million worldwide by the end of the century? But, the general feeling still seems to be, the end of the century . . . well, that is still a long way off, and who knows what will happen in between? Perhaps "they" will come up with a vaccine after all. It comes as something as a jolt to realize that the "end of the century" is now closer to us than the years we have suffered the AIDS pandemic to date.

Can anything stop the pandemic? At the moment I think not, given the difficulty in creating those breaks in the transmitting structures. In the Third World, what is there to produce those breaks in the backcloth? What will shred those structures in Africa, in India, in Indonesia, across those borders from Thailand into Burma, China, the refugee camps of Cambodia? Who, in the exhilaration of Mardi Gras, bothers with free condoms in Rio, São Paulo and other cities of Latin America celebrating a pre-lenten tradition in almost Bruegelesque fashion?

And what about the First World, with its own pockets of despair? In the crowded slums of people from Africa in French cities, in the crowded tenements of Northern Italy housing people from the Mezzogiorno seeking hope? And in the frayed and suspicious fabric of Eastern Europe whose walls once kept people in, but also helped to keep the HIV out – in all these examples and many others the HIV travels easily on backcloths of poverty and misery. But not exclusively so. Rich and poor, young and old, men and women . . . the HIV is no respecter of class, color, or anything else. Anyone will do.

The Third World has few, if any, really effective resources. The First World has some, but still seems incapable of thinking through how to apply them properly. Above all, it has the resource of money, of wealth, and once again appears to think that by throwing money at the problem it will somehow solve it. It does this in wartime, usually with success – depending, of course, on whose side you happen to be. It was tried with cancer, a massive campaign of money in the Nixon years that never even got close to the objectives it was shooting for at the time. The warlike terminology is hardly an accident. Funds to fight the AIDS epidemic are rising – tens of millions, followed by hundreds of millions, followed by billions – but only a proportion are properly directed in such a way that they can have the slightest telling impact. Enormous sums go to waste, thrown at the problems because no policy is properly informed at the moment by either science or compassion. Let me explain.

All evidence to the contrary, a contrariness that only points to our own humanity, we are still a people, a world, standing in the tradition of the Enlightenment, following all the informing threads of thinking that lead to that trust in the illuminating light of reason. The HIV, like any other devastating viral producer of mortal disease, will succumb to scientific investigation – *true* scientific investigation – or not at all. There is nothing else. This has very clear implications: public money

– the charitable, enforced and compulsory tithings of a people – must be allocated to virology and pharmacology, and those specialties like cell biology and immunology, where peer judgement can see a clear implication for stopping the effects of HIV, or at least retarding its effects, if the proposed research meets its goals. Even in these specialized life sciences the areas embraced by research are very wide, but I sense that real science is at work, and peer scrutiny and appraisal are rigorous. The criterion for allocating research funds must be utility, and only utility. Proposed research must be able to answer "yes" to the question: "If this research were funded, and if its goals are met, is it highly probable that we can use this knowledge to advance the day when we are released from the terrible effects of HIV?" If there is any doubt, any hemming and hawing, any obfuscation or Old Boy Waving of Hands in answering this question in the affirmative, someone, somewhere, must have the moral and political courage to deny the funds. Otherwise research on AIDS becomes just another opportunistic porkbarrel to support soft-money researchers in a style to which they have become accustomed.

I suggest that much, possibly most, of the research in the life sciences focusing upon HIV can pass such a sheep-and-goats test question. I suggest that most, the overwhelming majority, of the research programs in the human sciences cannot. Across the broad spectrum of the social and behavioral sciences, including epidemiology, most of the research is, quite literally, use-less. And I mean by that exactly what the emphasizing hyphen signifies; that it cannot be used, it has no utility, it cannot be applied in any conceivably plausible way to slow the epidemic, to stop one transmission, to save one human life. As someone whose own life has been devoted to the university and the best that it stands for, I have nothing against knowledge for knowledge's sake, and I honor insights that throw a genuine light of understanding into the dark corners – wherever these may be. But these efforts cannot be supported by public funds whose fundamental purpose is to stop the killing and comfort the dying. Research proposals cannot be hitched to the tumbrils carrying the dead to convert them into bandwagons for the living. This is not a scientific question, but a moral one.

As we have seen, most of the epidemiological studies have been stuck at their paradigmic deadcenters, and it seems that little will budge them. In turn, they, with their pretentious claims that their "theories" are outrunning the data, have given rise to a plethora of social and behavioral studies whose findings are always historical exercises, out of date and useless by the time they are published in journals of limited circulation that few peruse, let alone read. In 1992, who cares that we can be 95 percent certain that the HIV rate in prostitutes in São Paulo was 8.26

plus or minus 1.17 percent? That was in 1988, four years and a quarter of a million dollars in research funds ago. What on earth are you going to do with this knowledge, except use it to cut one more notch in your academic or professional *curriculum vitae*? Can you use it for education? No. Can you use it for planning? No – indeed, planning *what*? For forecasting? Forecasting what? Yet one more number down the time line that no one can do anything with? Could that quarter of a million been used in better, more *useful* ways, or for more compassionate purposes? And so, as the title of that 1987 book told us, The Band Play[s] On.

Are all the research programs in the so-called human sciences useless, doomed by failing to meet the criterion of utility? Perhaps not. Questions about the human dimensions of the epidemic can be posed, and often seek an urgent answer. But the posing may be much easier than the answering. As we saw, you cannot monitor teenage couplings for condom use, and many of the questions we have require testing large numbers of people for the HIV, perhaps a project impossible in a democratic society. Even now, all the states require the Wassermann test for syphilis before issuing a marriage license, but do not test for HIV on the grounds "it is too expensive." As we saw, it costs the US Armed Forces about $5.00. Too expensive for whom, to save the life of their partner and their children? Who is doing the testing, and in more ways than one "making a killing"?

In the universities, research requiring testing would never get past the ethical scrutiny of human subjects committees, whose charge is to protect people from any research by subterfuge. It is important to monitor, but long-term longitudinal studies in the human sciences tend to be rare, and the data tend to become historical long before they can serve as warning systems, early or late. Anyway, who needs more warning systems now? Since education is still all we have to fight the epidemic, any research genuinely focused on effective intervention deserves a hearing, but those educational coat-tails are long, and those trying to ride on them should have their proposals carefully inspected for that camel's nose under the tent. Those at the front line of AIDS education are shoulder-to-shoulder with those on the frontiers of virology and immunology, precisely because the products of their research have a clear and immediate use beyond the mere documentation of historical trivia. As for the rest of the social and behavioral sciences, our ability to fight the epidemic would be unaffected if most of their published findings were to disappear overnight. The next day no one would notice the disappearance of this pseudoscience, carried out not for useful knowledge, but for personal professional advancement.

That is the scientific side of the response to the pandemic: science as our only hope ultimately to stop the ravages of the HIV. The other side is the compassionate, just as demanding, and just as requiring of immense public funding. Perhaps in the long run (and I see no other "run" at the moment), this side may be more difficult than the scientific, because it demands of us a translation of caring into political will. This means engaging in the confrontations that will restore responsible government to the *demos* of a democracy, and not necessarily in the immediate and obvious political arenas, but along corridors of misplaced, but jealously guarded, power in many places. In the United States, and in varying degrees elsewhere (the problems of America are not unique), it means a redirection of funds to the compassionate tasks of caring for those in the terminal stages of affliction, and to the enlargement of healthcare systems already gravely overstressed. In 1989, when I lectured at the headquarters of one Public Health Region, I met a forthright doctor in a major metropolitan area who said that his region, and all the others across the country, had not even started to plan for the dying that could be seen coming out of the future, a future now here and still becoming. It was not that he was indifferent himself, for he obviously cared intensely, but simply that in the political climate of those days healthcare *planning* at the federal level was the equivalent of swimming against a riptide. After all, the private sector would take care of it – just like the private-for-profit jails that were then being considered, approved and built.

But tides turn, and taken at the flood . . .

Changing worlds, changing genres: a bibliographic essay

> It is to be hoped that this marks the end of the multitude of attitude studies that are adding little to the bank of international knowledge.
> *Dr Lorraine Sherr, St Mary's Hospital, London, in an editorial comment in* Current AIDS Literature, *4, 1991, p. 116*

"Pro bono publico" appears opposite the title page of this book and such is the firm intention. How, in all honesty, could it be anything else? But if that phrase is not totally redundant, it must contain more within it. And so it does. It also says that I usually find myself in a rather specialized world, governed quite properly by rather strict rules of writing, while wanting to reach out from time to time to a wider, perhaps more spacious world of public, rather than academic, communication. I happen to like that word "communication," and often reflect upon its roots in *com munus* – "with offering." At the etymological heart of communication lies a sharing of gifts. I feel that from time to time some of us in the academic world should try to share what we have found with others far beyond the circles of students that give so much purpose to our daily academic lives.

But in changing worlds, in making a move from that academic ivory tower towards the bustling streets and avenues of the surrounding public world, one also changes genres – ways and styles of writing, and in particular, ways of referencing. In the academic world – and as a perpetual student it has been my home for 40 years – close, attentive, sometimes obsessive referencing is obligatory on something approaching the moral high ground. Most academic writing tries to capture a sense of forward movement, no matter how hesitant, a change of perspective, no matter how small the shift in viewpoint. We try to see something not seen in quite the same way before and bring it into the light of greater understanding. Academic writing demands constant acknowledgement of the steps others have taken, of prior views, of previous ideas, and it documents and footnotes where those views and ideas may be found, when they were "made public," and who held them. No matter if many are simply genuflections, disclosing a certain amount of ritualistic behavior; at their best and most valuable they honor the intellectual ground upon which one stands at the moment of writing.

Writing in the larger public realm is different, although I would be the first to acknowledge it is always a difference of degree, of shading and movement along a spectrum of subtle gradations, rather than a difference of sharp breaks and clear distinctions. Perhaps the most public form of writing is journalism, containing within it its own sense of daily (*jour*) immediacy and ephemerality. It is certainly a distinct genre, and it makes its own severe demands of clarity and conciseness upon the writer. It is often qualified by academic people with the denigrating and despising adjective "mere." I happen not to share this disparaging view of daily writing, the best of which – and why talk about anything else? – is certainly "pro bono publico," perhaps the taproot of a democratic society, and to be honored as such. There is obviously fine and responsible journalism, and disgraceful and totally irresponsible journalism. It is, after all, a human endeavor. Why should we expect it to escape the predicament of human possibility? But let us not forget there is also fine and responsible academic writing, and academic writing so wretched, so tortuously obfuscating and essentially dishonest, that you wonder how the writer ever made it to the ground floor of the ivory tower.

In moving from the specialized academic world and genre of writing and referencing towards the more public world and genre of the responsible journalist I face not a dilemma, but certainly a difficulty in knowing how to resolve the question of referencing. Most readers in the wider public realm will become impatient with numerous notes and references constantly interrupting the flow of the text. Some may even suspect the writer of showing off when the footnotes are longer than the text itself. There are chapters in the book where this could have been achieved: some might have had over a hundred references to sources unavailable to even fine university and public libraries. After four years of specialized geographic research leading to numerous and purely academic publications and proposals, I face a pile of reference materials five to six feet high – not counting the books, but ranging from technical and academic journals to folders of newspaper clippings from Third World countries.

I recognize that some academic readers may not agree, but it seems to me that it ultimately comes down to a matter of trust. The reader either does, or does not, trust the writer. Once a book is launched on the sea of readership there is little if anything a writer can do. It is sink or swim. But the matter of trust goes further – much further – and I think it is important to take a moment or two to reflect upon it.

Not only is there an issue of trust between reader and writer, but also a matter of trust between a writer and his published sources. This always comes down to a matter of judgement, which is really another way of saying once again that it is a matter of trust. You judge a report reliable or unreliable depending on whether you trust it. Usually this means that you find it reasonable within the particular, but always larger context and set of conditions in which the report with its facts and figures is found. If a Minister of Health reports one year to the World Health Organization a lower figure of people with AIDS in his country than the year before, most people would find this suspicious, and tend not to trust it even though it appears in an official report. On the other hand, if a journalist interviews a doctor in a Third World hospital attached to a university, and then reports that 4.7 percent of the women coming for pre-natal care, and 7.4 percent of the young men of recruitable age are HIV infected, then I would be inclined to trust such figures even if they appeared in a "mere" newspaper, rather than waiting for them to appear in a peer-reviewed scientific journal four years later when the information would be of no use to anyone except the epidemiological historian. What do you think? How would you weigh it up in your own judgement? Does it sound reasonable? Would you trust it?

Academic people, while always doubting for the sake of the truth, tend to be a trusting lot, because science, scientific research, and scientific reporting are based on trust. Unfortunately, publication in a peer-reviewed scientific journal is no guarantee that trust has been properly placed. Most articles in medically-related journals have multiple authors, sometimes up to ten or a dozen. You know perfectly well that many of them had little if anything to do with the research reported, that some are administrative bureaucrats, while others were listed for their possible halo effect. One researcher has over 900 scientific, peer-reviewed articles to his credit because he insists that his name be included on any paper making use of "his" virus in other, totally independent research projects. Even worse, data is sometimes forged: spots are painted on mice and reported as tumors, or photographs of someone else's virus get published "by mistake." Notebooks get lost, and many reported studies are found to be nonreplicable – if replication is ever tried, which often it is not. In the United States alone, the national institutions responsible for funding and overseeing scientific research are constantly investigating and exposing fraudulent findings, some of them in the most prestigious research institutes and universities in the country. The imprimatur of peer-review in scientific journals, while it undoubtedly helps, is no guarantee that the facts and figures are trustworthy. Whether in the world of the academic, or the world of the journalist, the issue of trust will not go away.

There is one further issue of trust to make explicit, the issue of trust between a writer and his verbal sources. In moving in that intermediate zone between academic research and reports of daily immediacy in the public realm, I sometimes found myself in the position of the journalist who comes across important information but is obliged to attribute it to a "spokesman." One does this usually because to reveal the identity of the source would place it (and the person or persons behind it) in jeopardy. Several times I have come across people in the lower ranks of a bureaucracy who have felt that the issue of confidentiality was being misused by senior officials, that data and information were being unfairly and unreasonably withheld, and that the responsible course was to reveal it. One academic reviewer of this manuscript asked "How does he know Arab sheiks charter planes for sex tours to Bangkok?" (page 92), implying either that I should have footnoted *The Journal of Immoral Purposes*, vol. xxi, 1986, p. 236, or that I had made the whole thing up out of thin air. Now it happens that this fact comes from a friend of 30 years standing whose veracity I trust. I will not disclose such a source, or anything about him which could conceivably lead to his identification, because his long and distinguished career would be placed in jeopardy. Whether you trust my judgement, here as elsewhere, is for you to decide.

As one of the most serious pandemics ever faced by the human family, the spread of HIV/AIDS has generated a colossal literature ranging across many fields of the medical, biological and human sciences. It has also influenced much humane writing and literature, and been the subject of almost daily bureaucratic reports and public sources of information, such as newspapers, television and weekly journals. In the late 1980s, one of my students and I tried to monitor and record the accelerating upward curve of major newspaper reports in the United States alone. After about 1988 it became pointless: no matter what the area of writing, journalistic or scientific, the explosion of writing followed the same explosion of the virus across the world, even if much of the reporting was repetitive, redundant, or too often trivial. There are now not only specialized scientific journals devoted solely to AIDS research, but monthly reviews of scientific AIDS literature. By 1991, one of them, *Current AIDS Literature*, was reporting 951 articles in the first three months of that year, culled from 71 medical and related journals, all but 11 of them in English. Even so, there is virtually nothing in this huge and burgeoning flow on the geography and geographic diffusion of the HIV. Articles with titles naming specific places – towns, cities, regions, nations – are precisely that: time and space specific, usually reporting information such as rates of infection about specific groups of people, with little sense of the geographical dynamics that might be involved. Simply to scan and record such information in its (x,y,t) coordinates is an almost impossible task, and to place such information properly into its spatial and temporal coordinates, i.e. to create an AIDS Geographic Information System (GIS), would require a massively funded monitoring team. This has never been done.

Given the plethora of AIDS and related publications, I can only give you an introduction to this massive literature. I will start with general sources of information that I have found useful, limiting the list that may be reasonably available to an English-speaking person with access to a university or public library with inter-library loan facilities.

The World Health Organization (WHO) publishes a series of reports on various aspects of the global response, and the WHO bibliography already lists hundreds of references. One might wish to start with:

Global Programme on AIDS, a WHO annual report, WHO Publications, Geneva.

WHO also publishes the:

Weekly Epidemiological Record, WHO Publications, Geneva.

These contain figures officially reported, which the WHO is obliged to accept at face value. One of the most wide-ranging sources of information is the bimonthly Panos Institute publication:

WorldAIDS, Panos Institute, London, Washington, DC, Paris, Budapest, whose articles and reports are documented as to the original source. Panos also publishes books devoted to the pandemic in their series *Panos Dossier*, often revising these to keep the information up to date, for example:

AIDS and the Third World, Panos Institute, London, Washington, DC, Paris, Budapest, 1988 (third edition);

Blaming Others: Prejudice, Race and Worldwide AIDS, Panos Institute, London, Washington, DC, Paris, Budapest, 1988;

The Third Epidemic: Repercussions of the Fear of AIDS, Panos Institute, London, Washington, DC, Paris, Budapest, 1989.

All three books were published in conjunction with the Norwegian Red Cross, and contain detailed documentation as to the original sources. Other reviews and bulletins include:

AIDS and Society, published by the African-Caribbean Institute,

AIDS Updates.

The former is mainly devoted to reports from Africa, the Caribbean and Latin America, while the latter contains review articles, many of which may document several hundred reports, articles and reviews.

The most complete review in terms of a broad spectrum of research in the medical, epidemiological and human sciences is:

Current AIDS Literature, monthly publication of the Bureau of Hygiene and Tropical Diseases, Current Science, London and Philadelphia.

Most research on AIDS is published in traditional journals on medicine, virology, epidemiology, etc., but three journals are now devoted entirely to AIDS research:

AIDS

AIDS Research and Human Retroviruses

Journal of Acquired Immune Deficiency Syndrome.

Most articles in these journals are highly specialized reports, but more general epidemiological information is frequently included, often as short reports or correspondence.

More general sources of information are found from time to time in:

World•Watch, Worldwatch Institute, Washington, DC,

a bi-monthly publication of the Worldwatch Institute that takes up many global problems, while the internationally-known journals *Science*, *Nature* and *Scientific American* occasionally devote whole issues to the AIDS pandemic in articles generally accessible to the non-specialist, as well as publishing highly specialized reports. In particular:

Science, 239, 1988, pp. 533–696

Scientific American, 259, 1988, pp. 40–152.

Major newspapers such as *The Times* and the *Guardian* (UK), and the *New York Times*, *Christian Science Monitor*, and *Washington Post* (USA), etc. should also be consulted. They, and other newspapers of repute, often run special articles on AIDS/HIV, generally well-researched.

Many national medical associations have produced publications to inform both medical and lay people, for example:

British Medical Journal, *ABC of AIDS*, London, 1988,

while in the United States, the Centers for Disease Control regularly issue:

HIV/AIDS Surveillance, US Department of Health and Human Services, Atlanta, GA

whose updated tabular material is also regularly copied and reported in the journal *AIDS*. The weekly:

Morbidity and Mortality Weekly Report, Centers for Disease Control, Atlanta, GA

also frequently carries reports on HIV/AIDS. In the United States, virtually all state health departments now issue regular (monthly, bi-monthly, quarterly) AIDS reports, most of which now break down state totals into county values. Many contain articles on the epidemic within the state boundaries (or even beyond), and I have found the reports of California, Washington, Illinois and New York particularly valuable. Hawaii's AIDS Task Group is willing to share its *Minutes of Meetings* (1951 East-West Road, Honolulu, HI 96522), and I have found their caring response, openness and wide-ranging discussions outstanding.

A reader whose curiosity has been piqued by materials in individual chapters may wish to follow up some of them by consulting the references I have listed under each of the chapter headings below. I have tried to include references which themselves contain copious and detailed bibliographic information, so that the specialist or layperson who wishes to dig deeply may follow these in an almost chain-like bibliographic reaction.

Prologue: new plagues for old

There are whole bibliographies on the history of previous epidemics and pandemics, but a few items will lead a reader into an extraordinarily rich historical literature. A short, but geographically aware article on the Black Death is:

W. Langer, "The Black Death," *Scientific American*, 210, 1964, pp. 114–121;

while the syphilis story is told in:

C. Quétel, *History of Syphilis*, Johns Hopkins University Press, Baltimore, 1990.

Thoroughly geographical perspectives on the great influenza pandemic of 1918–19, and subsequent epidemics in the United States, are given in:

G. Pyle, *The Diffusion of Influenza: Patterns and Paradigms*, Rowman and Littlefield, 1986;

K. Patterson and G. Pyle, "The Geography and Mortality of the 1918 Influenza Pandemic," *Bulletin of the History of Medicine*, 65, 1991, pp. 4–21.

1 The killer: HIV and what it does

The identification of the HIV viruses, and the explication of their details and mechanisms, is one of the great stories of virology, producing a huge and highly specialized literature that is still growing. As a non-specialist, I found the following more general articles extremely helpful in trying to limn the structure of the viruses and their related forms:

R. Gallo, "The First Human Retrovirus," *Scientific American*, 224, 1986, pp. 88–98;

Z. Rosenberg, A. Fauchi, "Inside the AIDS Virus," *New Scientist*, 10 February 1987, pp. 51–4;

M. Koch, "The Anatomy of the Virus," *New Scientist*, 26 March 1987, pp. 46–51;

P. Nada, "AIDS Viruses of Animals and Man," *Los Alamos Science*, 18, 1989; pp. 54–89.

The epidemiological history up to about 1989 is told by one of the great historians of medicine:

M. Grmek, *History of AIDS: Emergence and Origin of a Modern Pandemic*, Princeton University Press, Princeton, 1990.

2 The origins of HIV: closing an open question?

Many highly specialized reports in medical journals have contributed to the question of HIV's origins. Useful starting points to follow the hunt include:

F. Barin, S. M'Baut, F. Denis, "Serological evidence for a virus related to simian T-lymphotropic retrovirus III in residents of West Africa," *The Lancet*, 2, 1985, pp. 1387–9;

I. Bygbjer, "AIDS in a Danish Surgeon (Zaïre, 1976)," *The Lancet*, 1, 1983, p. 925;

J. Desmyter, I. Surmont, *et al.*, "Origins of AIDS," *British Medical Journal*, 293, 1986, p. 1308;

M. Essex, P. Kanki, "The origins of the AIDS virus," *Scientific American*, 259, 1988, pp. 64–71;

J. Gonzales, M. Georges-Courbot, *et al.*, "True HIV-1 infection in a pygmy," *The Lancet*, 1, 1987, p. 1490;

A. Nemeth, S. Bygdeman, *et al.*, "Early case of acquired immunodeficiency syndrome in a child from Zaïre," *Sexually Transmitted Diseases*, April–June, 1986, pp. 111–13;

N. Nzilambi, K. DeCock, *et al.*, "The prevalence of infection with a human immunodeficiency virus over a 10-year period in rural Zaïre," *New England Journal of Medicine*, 318, 1988, pp. 276–9;

P. Brown, "Earliest AIDS case may offer clues to virus," *New Scientist*, 14 July, 1990, p. 34, reporting on a communication to *The Lancet*, 336, 1990, p. 51;

A. Nahmias, *et al.*, "Evidence for human infection with an HTLV-III/LAV-like virus in Central Africa," *The Lancet*, 1, 1986, p. 1279–80;

R. Sher, *et al.*, "Seroepidemiology of HIV in Africa from 1970 to 1974, *New England Journal of Medicine*, 317, 1987, pp. 450–7;

G. Corbitt, *et al.*, "HIV infection in Manchester, 1959', *The Lancet*, 336, 1990, p. 51;

J. Sonnet, "Early AIDS cases originating from Zaïre and Burundi (1962–76)," *Scandinavian Journal of Infectious Diseases*, 19, 1987, pp. 511–17;

S. Frøland, *et al.*, "HIV-1 infection in a Norwegian family before 1970," *The Lancet*, 8598, 1988, p. 1344;

D. Huminer, J. Rosenfeld, S. Pitlik, "AIDS in the pre-AIDS era," *Review of Infectious Diseases*, 9, 1987, pp. 1102–8.

Anthropological speculations as to possible transmission mechanisms between animal and human populations include:

A. Kashamura, *Famille, sexualité et culture: Essai sur les moeurs sexuelles et les cultures des peuples des Grands Lacs Africans*, Payot, Paris, 1973.

While evidence from geographic modeling at the global level comes from:

A. Fahault, A-J. Valleron, "The role of air transport in the global spread of HIV infection," paper available from Unité de Recherches Biomathématiques et Biostatistiques, Université de Paris VII.

Their global airline model is based on:

L. Rvachev, I. Longini, "A mathematical model for the global spread of influenza," *Mathematical Biosciences*, 75, 1985, pp. 3–22;

I. Longini, P. Fine, S. Thacker, "Predicting the global spread of new infectious agents," *American Journal of Epidemiology*, 123, 1986, pp. 383–91;

Genetic variations in HIV and SIV include:

B. Hahn, *et al.*, "Genetic variations in HTLV–III/LAV over time in patients with AIDS or at risk of AIDS," *Science*, 232, 1986, pp. 1548–53;

T. Huet, *et al.*, "Genetic organization of a chimpanzee lentivirus related to HIV-1," *Nature*, 345, 1990, pp. 356–9;

N. Letvin, M. Daniel, *et al.*, "An HIV related virus from Macaques," in J. Gluckman (ed.), *Acquired Immunodeficiency Syndrome*, Elsevier, Amsterdam, 1987, pp. 71–4;

R. Desrosiers, "A finger on the missing link," *Nature*, 345, 1990, pp. 288–9;

C. Joyce, "Viral mutation rate alarms AIDS researchers," *New Scientist*, 4, June 1987, p. 28;

M. McClure, "AIDS and the monkey puzzle," *New Scientist*, 25 March 1989, pp. 46–52;

P. Brown, "The strains of the HIV war," *New Scientist*, 25 May 1991, pp. 20–21.

3 The thin tendrils of effects

To document the effects of HIV on human life would require a bibliography of literally thousands of references, itself indicative of the numerous and widespread implications. References here are starting points and can only indicate often large, sometimes enormous, literatures. The matter of burn-out among young nurses and doctors has had widespread discussion, including:

L. Bennett, "Burnout – what's burnout?," *WorldAIDS*, 9, 1990, p. 2;

M. Helqvist (ed.), *Working with AIDS: A Resource Guide for Mental Health Professionals*, AIDS Health Project, University of California at San Francisco, San Francisco, 1987.

The appalling decisions leading to massive HIV transmission by the blood industry in the United States are described by Shilts (see p. 221 below), but France has also suffered traumatically from the same problem, resulting in legal prosecutions. In India, the situation is discussed in:

O. Sattaur, "India wakes up to AIDS," *New Scientist*, 1 November 1991, pp. 25–9.

The financial opportunities accruing to the international arms trade are brought out in:

A. Venter, "AIDS: Its strategic consequences in Black Africa," *International Defense Review*, 21, 1988, pp. 357–9.

Any financial analyst can pull portfolios of "AIDS stocks" for potential investors in the pandemic, including pharmaceutical and condom manufacturers. Economic subsidies are discussed in:

O. Sattaur, "Condom boom brings new life to rubber substitute," *New Scientist*, 24 March 1990, p. 32.

Some of the difficulties experienced by the insurance industry are engaged in:

B. Schatz, "The AIDS insurance crisis: underwriting or overreaching?," *Harvard Law Review*, 100, 1987, pp. 1782–1804;

D. Stone, "The rhetoric of insurance law: the debate over AIDS testing," *Law and Social Inquiry*, 15, 1990, pp. 385–407;

including many legal aspects. The reader should be aware that hundreds of law cases are discussed, and a bibliography would run into thousands of references. A feel for some of the major issues may be found in:

R. Wasson, "AIDS discrimination under federal, state and local law after Arline," *Florida State University Law Review*, 15, 1987, pp. 221–78;

L. Gostin, W. Curran, M. Clark, "The case against compulsory testing in controlling AIDS: testing, screening and reporting," *American Journal of Law and Medicine*, 12, 1987, pp. 7–53;

D. Merritt, "Communicable disease and constitutional law: controlling AIDS," *New York University Law Review*, 61, 1986, pp. 739–99;

while an excellent review of the relations between law and medicine will be found in:

L. Gostin, "The AIDS litigation project: a national review of court and human rights commission decisions: Part I, the social impact of AIDS; Part II, discrimination," *Journal of the American Medical Association*, 263, 1990, pp. 1961–70 and 2086–93.

The tragedy of AIDS in the world of the arts is documented in many plays, and increasingly in dance performances. Again Shilts (see p. 221 below), discusses many of these responses, but a poignant work is:

M. Klein (ed.), *Poets for Life: Seventy Six Poets Respond to AIDS*, Crown Publishers, New York, 1989.

while the courage of individual response is found in such works as:

E. Dreuilhe, *Mortal Embrace: Living with AIDS*, Hill and Wang, New York, 1988.

4 Sex on a set: a backcloth for disaster

While the basic ideas of relations structuring sets to form backcloths allowing for the transmission of traffic are intuitively simple, mathematical treatments are also available for the technically-minded reader. These would include both popular works such as:

R. Atkin, *Multidimensional Man: Can Man Live in 3-Dimensional Space?* Penguin Books, Harmondsworth, 1981;

and more technical pieces such as:

R. Atkin, *Mathematical Structure in Human Affairs*, Heinemann Educational Books, London, 1974;

and the journal

Planning and Design: Special Issue on Q-Analysis, 10, 1983, Pion Ltd, London.

Traditional network analysis applied to epidemiological tracing in the context of the AIDS pandemic appears in:

D. Auerbach, *et al.*, "Cluster of cases of the acquired immune deficiency syndrome: patients linked by sexual contact, *The American Journal of Medicine*, 76, 1984,

while an imaginative statistical use of network analysis is made in:

E. Laumann, *et al.*, "Monitoring AIDS and other rare population events: A network approach," paper presented at the Sunbelt Conference of the International Society for Social Network Analysis, Tampa, Florida, January 1991, pp. 51;

E. Laumann, *et al.*, "Monitoring the AIDS epidemic: a network approach," *Science*, 244, 1989, p. 1186.

A general, but advanced work on network analysis is:

B. Wellman, S. Berkowitz, *Social Structures: A Network Approach*, Cambridge University Press, 1988;

while elementary discussions of functions used to describe epidemics over time may be found in:

R. Abler, J. Adams, P. Gould, *Spatial Organization: The Geographer's View of the World*, Prentice Hall, Englewood Cliffs, 1971.

5 Transmission break: a geography of the condom

I know of no geographical research on the condom, and virtually nothing on the geography of sexual relations, although the fine and pioneering work by a geographer:

R. Symanski, *The Immoral Landscape: Female Prostitution in Western Societies*, Butterworths, Toronto, 1981;

is increasingly referenced by other human scientists. Numerous short reports on condom use and propagation are given in almost all issues of *WorldAIDS*, while many articles and reports in the "AIDS Monitor" of *New Scientist* deal with condom use. These include:

B. Spencer, "An open French letter," *New Scientist*, 9, November 1991, pp. 58–9;

P. Brown, "Africa's growing AIDS crisis," *New Scientist*, 17 November 1990, pp. 38–41;

S. Kingman, S. Connor, "The answer is still a condom," *New Scientist*, 23 June 1988, pp. 33–6.

For a scholarly, yet humanly moving article, see:

C. Taylor, "Condoms and cosmology: the 'fractal' person and sexual risk in Rwanda," *Social Science and Medicine*, 9, 1990, pp. 1023–28,

while changing social and economic conditions are outlined in:

J. Gamson, "Rubber wars: struggles over the condom in the United States," *Journal of the History of Sexology*, 1, 1990, pp. 268–82.

6 How things spread: hierarchical jumps and spatial contagion

There is a very large literature here, ranging from elementary textbook discussions to advanced works on geographic diffusion theory. Two elementary but useful starting points might be my chapters "Human Contacts in Space and Time," and "Geography and Medicine: An Old Partnership," in:

P. Gould, *The Geographer at Work*, Routledge, London and New York, 1984,

as well as the chapter "Spatial Diffusion" in Abler, Adams and Gould (above). A more advanced discussion will be found in:

A. Cliff, P. Haggett, "Spatial aspects of epidemic control," *Progress in Human Geography*, 13, 1989, pp. 315–47.

which contains a useful introductory bibliography.

A definitive work is:

A. Cliff, P. Haggett, *Atlas of Disease Distributions: Analytical Approaches to Disease Data*, Blackwell, Oxford, 1988.

In the specific context of the AIDS epidemic, geographic aspects are approached in:

G. Shannon, G. Pyle, R. Bashshur, *The Geography of AIDS*, Guilford Press, New York, 1990;

G. Shannon, G. Pyle, "The origin and diffusion of AIDS: a view from medical geography," *Annals of the Association of American Geographers*, 79, 1989, pp. 1–24;

L. Gardner, J. Brundage, *et al.*, "Spatial diffusion of the human immunodeficiency virus infection epidemic in the United States, 1985–87," *Annals of the Association of American Geographers*, 79, 1989, pp. 25–43.

W. Wood, "AIDS north and south," *Professional Geographer*, 40, 1988, pp. 266–79,

while a fine and extraordinarily detailed work on the geography and many other aspects of the HIV/AIDS pandemic is:

M. Smallman-Raynor, A. Cliff, P. Haggett, *Atlas of AIDS*, Blackwell, Oxford, 1992.

7 Africa: a continent in catastrophe

Probably more has been written at both the highly technical, as well as the more general level about HIV/AIDS in Africa than any other continent. Newspapers in Africa and around the world comment almost daily on some aspect or another. Books range from social science reports with massive bibliographies and research guides, such as:

N. Miller, R. Rockwell (eds.), *AIDS in Africa: The Social and Policy Impact*, Edwin Mellen Press, Lewistown, 1988;

to detailed, personal and ancedotal accounts by reporters such as:

E. Hooper, *Slim: A Reporter's Own Story of AIDS in East Africa*, Bodley Head, London, 1990.

An enormous medical literature is culled for rates of HIV and AIDS and compiled as *Research Notes* by the US Bureau of the Census, which ships an updated disk to the WHO in Geneva each month. Such reports include:

Seroprevalence of HIV in Africa: Winter 1990, Center for International Research, US Bureau of the Census, Washington, DC, 1990;

Trends and Patterns of HIV/AIDS Infections in Selected Developing Countries: Country Profiles, Health Studies Branch, US Bureau of the Census, Washington, DC, 1992;

Recent HIV Seroprevalence Levels by Country: April 1992, Heath Studies Branch, US Bureau of the Census, Washington, DC, 1992.

References for this chapter would also include several hundred newspaper clippings from West, East, Central and South Africa, most, but not all of them, from English-speaking areas, as well as educational pamphlets and posters. On perceptions of death see:

B. Fleming, "Another way of dying," *The Nation*, April 2, 1990, pp. 446–50.

The satellite detection of vegetation changes indicating the desertion of villages in AIDS-stricken areas in East and Central Africa was noted in the *Minutes of Meeting*, February 23, 1990, of the Hawaii AIDS Task Group, and was "confirmed" by a clipping from a South African newspaper written by T. Nicholson, "The big AIDS deception." I have been unable to trace this account further, but I see no reason why rapid growth of bush in abandoned regions should not be detectable.

The impact on agricultural production has been thoughtfully discussed in:

T. Barnett, P. Blaikie, "AIDS and food production in East and Central Africa," *Food Policy*, February 1989, pp. 2–6;

A. Barnett, *et al.*, "The impact of AIDS on the food production systems and rural economies of East and Central Africa over the next ten years," paper, Conference on the Global Impact of AIDS, London, 8–10 March, 1988;

while the impact on mining has been examined in:

B. Nkowane, "Implications of HIV and AIDS for primary industry: mining, a case study for Zambia," paper, Conference on the Global Impact of AIDS, London, 8–10 March, 1988.

Political responses and commentary include:

A. Fortin, "The politics of AIDS in Kenya," *Third World Quarterly*, 9, 1987, 906–19;

while some of the many obstacles are outlined in:

B. N'Galy, S. Bertozzi, R. Ryder, "Obstacles to the optimal management of HIV infection/AIDS in Africa," *Journal of Acquired Immune Deficiency Syndromes*, 3, 1990, 430–7.

The social context is reviewed (with an excellent bibliography) by:

J. Caldwell, P. Caldwell, P. Quiggin, "The social context of AIDS in sub-Saharan Africa," *Population and Development Review*, 15, 1989, pp. 185–234;

while the effects of war are analyzed in:

M. Smallman-Raynor, A. Cliff, "Civil war and the spread of AIDS in Central Africa," *Epidemiology of Infectious Diseases*, 107, 1991, pp. 69–80.

8 Thailand: how to optimize an epidemic

Major sources of information, including rates in selected populations and regions, are published frequently in the English-language press of Thailand including *The Bangkok Post*, *The Nation*, and the *Northern Monthly*. These reports by journalists appear frequently as references in scholarly articles in peer-reviewed journals. I consider them highly reliable. It is also clear from the language used that many larger review articles in *The Far Eastern Economic Review*, *New Scientist*, *WorldAIDS*, *The Economist*, the *New York Times*, etc. utilize such reports. I have used, literally, hundreds of clippings to piece together the estimated map and the accompanying narrative.

The question of women in Thailand being exploited as sex workers aroused international concern with an International Labour Organization publication:

P. Phongpaichit, *From Peasant Girls to Bangkok Masseuses*, International Labour Office, Geneva, 1982,

which contains references to an earlier literature in Thai. However, many valuable papers by Thai researchers appear in English, for example:

V. Poshyachinda, "Heroin in Thailand," Drug Dependence Research Center, Chulalongkorn University, Bangkok, 1982;

V. Poshyachinda, "Recent changes in needle sharing among intravenous injecting heroin abusers in Thailand," Institute of Health Research, Chulalongkorn University, Bangkok, 1989;

V. Poshyachinda, "Illegal opiate consumption in Thailand's population and use pattern," Drug Dependence Research Unit, Chulalongkorn University, Bangkok, 1989,

all of which also contain references in English.

The social aspects are analyzed in:

N. Ford, S. Koetsawang, "The socio-cultural context of the transmission of HIV in Thailand," *Social Science and Medicine*, 33, 1991, pp. 405–14;

N. Ford, "The social context of the emergence of HIV in Thailand," *Journal of Population and Social Studies*, 2, 1990, pp. 223–37;

E. Cohen, "Sensuality and venality in Bangkok: the dynamics of cross-cultural mapping of prostitution," *Deviant Behavior*, 8, 1987, pp. 223–34;

S. Karel, B. Robey, "AIDS in Asia and the Pacific," *Asian and Pacific Population Forum*, 2, 1988, pp. 23–9,

while discussions of the international aspects of the sex trade involving Thailand include:

E. Cohen, "Tourism and AIDS in Thailand," *Annals of Tourism Research*, 15, 1988, pp. 467–86;

N. Ford, "Sex tourism to Thailand, *Isis*," 13, 1979, pp. 9–12, translated from the Dutch in *Onze Wereld* NOVIB, Amaliastraat 5–7, The Hague;

G. Ohi, I. Kai, *et al.*, "AIDS prevention in Japan and its cost–benefit aspects," *Health Policy*, 8, 1987, pp. 17–27;

S. Karel, B. Robey, "AIDS in Asia and the Pacific," *Asian and Pacific Population Forum*, 2, 1988. pp. 23–9.

Intervention efforts are discussed in:

M. Muecke, "The AIDS prevention dilemma in Thailand," *Asian and Pacific Population Forum*, 4, 1990, pp. 1–8, pp. 21–7;

M. Timm, "Condom cabaret in Bangkok," *New Internationalist*, November 1989, pp. 20–21,

and they are also touched upon in the interview with Dr Werasit Sittitrai in the *Minutes of Meeting*, June 28, 1991, by the Hawaii AIDS Task Group.

Information about land speculation and development in the north appears in:

S. Ekachai, "Under the speculator's hammer," in *Behind the Smile*, Thai Development Support Committee, Post Publishing, Bangkok, 1991, pp. 131–180,

while the status of women, and the attempts to empower women are discussed in:

D. and P. Tantiwiramanond, "Emergency home," in *By Women, For Women*, Institute of South East Asian Studies, Singapore, 1991, pp. 77–140.

Official medical programs and statistics are published by Thailand's Ministry of Health and WHO-associated bodies, for example:

M. Viravaidyan, "AIDS in the 1990s: meeting the challenge," Prime Minister's Office, Bangkok, 1991;

Ministry of Public Health, *Medium Term Programme Review*, WHO–MPH, Bangkok, 1991,

while unpublished materials are frequently reported by the epidemiological bibliographic service of Walter Reed Hospital, Washington, DC, for example:

K. Wagchusak, *et al.*, "Trends of HIV spreading in Thailand detected by national sentinel serosurveillance," MPH, Bangkok, 1992, in EPI Studies, Division of Preventive Medicine, Walter Reed Hospital, Washington, DC.

9 America: leaks in the system

No published work references the research underlying this chapter, except HIV/AIDS reports of individual states, and the academic publication below. All rates and cumulative values underpinning the maps were computed by Joseph Kabel:

J. Kabel, *A Geographic Perspective on AIDS in the United States: Past, Present and Future*, PhD Dissertation, University Park, PA, 1992.

All maps were constructed by me, and produced for publication here by Joseph Kabel, Ralph Heidl and William Holliday. The research was unsupported by any federal or state agency. Preliminary colored versions of the five-map sequence showing the diffusion of AIDS in the United States from 1982 to 1990 were shown at the International Congress on AIDS, Amsterdam, 1992, by Dr M. Fullilove, and have since appeared in *Time* August 31, 1992, p. 20.

10 The Bronx: poverty, crack and HIV

The publications from the long-term research program of Drs Rodrick and Deborah Wallace are definitive here. On fire, fire control, and spatially contagious burn-out, see:

R. Wallace and D. Wallace, *Studies in the Collapse of Fire Services in New York City 1972–1976: The Impact of Pseudo-science in Public Policy*, University Press of America, Washington, DC, 1977;

R. Wallace and D. Wallace, "Urban fire as an unstablized parasite: the 1976–1978 outbreak in Bushwick, Brooklyn," *Environment and Planning A*, 15, 1983, pp. 207–26;

R. Wallace and D. Wallace, "Structural fire as an urban parasite: population dependence of structural fire in New York City and its implications, *Environment and Planning A*, 16, 1984, pp. 249–60;

R. Wallace, "Contagion and incubation in New York City structural fires, 1964–1976," *Human Ecology*, 6, 1978, pp. 423–33;

R. Wallace, "The New York City fire epidemic as a toxic phenomenon," *International Archives of Occupational and Environmental Health*, 50, 1982, pp. 31–51;

D. Wallace, "An index of fire control service adequacy and its application on four neighborhoods of Manhattan," *Fire Technology*, August, 1982, pp. 170–84;

D. Wallace and R. Wallace, "A warning for European cities: the burning down of New York City, its causes and its impacts," *Anthropos*, 12, 1990, 256–72;

R. Wallace, "Fire service productivity and the New York City fire crisis," *Human Ecology*, 9, 1981, pp. 433–64.

The social consequences are brought out in:

R. Wallace, "A synergism of plagues: 'planned shrinkage,' contagious housing destruction and AIDS in the Bronx," *Environmental Research*, 47, 1988, pp. 1–33;

R. Wallace and D. Wallace, "Origins of public health collapse in New

York City: the dynamics of planned shrinkage, contagious urban decay and social disintegration," *Bulletin of the New York Academy of Medicine*, 66, 1990, pp. 391–434;

R. Wallace, "Expanding coupled shock fronts of urban decay and criminal behavior: how US cities are becoming 'hollowed out'," *Journal of Quantitative Criminology*, 7, 1991., pp. 333–55;

R. Wallace, "Homelessness, contagious housing destruction and municipal bus service cuts in New York City: I. Demographics of a housing deficit," *Environment and Planning A*, 21, 1989, pp. 1585–1603;

R. Wallace, "Homelessness, contagious housing destruction and municipal bus service cuts in New York City: II. Dynamics of a housing famine," *Environment and Planning A*, 22, 1990, pp. 5–15;

R. Wallace, "Social disintegration and the spread of AIDS: meltdown of sociogeographic structure in urban minority neighborhoods," *Social Science and Medicine*, 32, 1991.

The further consequences of social decay for the diffusion of HIV and the increase in AIDS are analyzed in:

R. Wallace, "Urban desertification, public health and public order: planned shrinkage, violent death, substance abuse and AIDS in the Bronx," *Social Science and Medicine*, 31, 1990, pp. 801–13;

R. Wallace and M. Fullilove, "AIDS deaths in the Bronx 1983–1988: spatio-temporal analysis from a sociogeographic perspective," *Environment and Planning A*, 1991, vol. 23, pp. 1701–23;

R. Wallace, J. Pittman, "Recurrence of contagious urban desertification and the social thanatology of New York City," available from the authors, New York Psychiatric Institute, Box 47, 722 West 168th Street, New York City, 10032;

R. Wallace, "Social disintegration and the spread of AIDS: thresholds for propagation along sociogeographic networks," available from the author, PISCS, 549 West 123 Street, New York City, 10027.

Estimates of infection in the Bronx, and the rise of tuberculosis, are recorded in:

E. Drucker, *et al.*, "Increasing rate of pneumonia hospitalizations in the Bronx: a sentinel indicator for human immunodeficiency virus," available from the authors, Montifiore Medical Center, 111 East 210th Street, New York City, 10467;

E. Drucker, S. Vermund, "Estimating population prevalence of human immunodeficiency virus infection in urban areas with high rates of intravenous drug use: a model of the Bronx in 1988," *American Journal of Epidemiology*, 130, 1989, pp. 133–42.

11 The response: how many bureaucrats can dance on the head of a pin?

Early responses to the AIDS pandemic were recorded in:

R. Shilts, *And the Band Played On*, St. Martin's Press, New York, 1987, in my opinion the finest account yet to appear, detailing the course of the AIDS crisis from the most personal to the highest bureaucratic levels.

The recommendations of the "White House Conference" appeared as:

Office of Science and Technology, *A National Effort to Model AIDS Epidemiology*, Executive Office of the President, Washington, DC, 1988.

As noted above, updated data sets on the pandemic are compiled monthly by the US Bureau of the Census and sent to the WHO in Geneva.

12 Time but no space: the failure of a paradigm

The quotation of Martin Heidegger regarding the way in which a particular theoretical perspective closes off the possibility of asking alternative questions is contained in:

M. Heidegger, "Science and Reflection," in *The Question Concerning Technology and Other Essays*, Harper Colophon Books, New York, 1977, p. 169.

Articles and books typical of what I have called the differential paradigm approach are:

R. Anderson, R. May, "Understanding the AIDS pandemic," *Scientific American*, May 1992, pp. 58–66;

R. Anderson, R. May, *Infectious Diseases of Humans: Dynamics and Control*, Oxford University Press, Oxford 1991;

R. Anderson, R. May, *et al.*, "The spread of HIV-1 in Africa: sexual contact patterns and predicted demographic impact of AIDS," *Nature*, 352, 1991, pp. 581–9;

R. Anderson, "The role of mathematical models in the study of HIV transmission and the epidemiology of AIDS," *Journal of Acquired Immune Deficiency Syndromes*, 1, 1988, pp. 241–56,

but they run into the hundreds.

Modeling the increase of AIDS in New York City was undertaken in:

R. Anderson, R. May, *et al.*, "Drugs, sex and HIV: a mathematical model for New York City," available from the authors at Imperial College of Science and Technology, Prince Consort Road, London, SW7 5BB, UK,

and the continuing strength of the paradigm to the end of 1992 is demonstrated in:

P. Rosenberg, *et al.*, "Population-based monitoring of an urban HIV/AIDS epidemic," *Journal of the American Medical Association* 268, 1992, pp. 495–503,

despite a throwaway paragraph alluding to the geographic dimensions of the epidemic at the end of the article by the 13 authors.

Backcasting estimates are undertaken in:

P. Rosenberg, M. Gail, "Backcalculation of flexible linear models of the human immunodeficiency virus infection curve," *Applied Statistics*, 40, 1991, pp. 1–15;

M. Gail, P. Rosenberg, J. Goedert, "Therapy may explain recent deficits in AIDS medicine," *Journal of Acquired Immune Deficiency Syndromes*, 3, 1990, pp. 296–306;

P. Rosenberg, M. Gail, "Uncertainty in estimates of HIV prevalence derived by backcalculation," *Annals of Epidemiology*, 1, 1990, pp. 105–115;

while the policy-making, not to say humane, implications of such temporal forecasting are brought out in:

J. Osborn, "Policy implications of the AIDS deficit," *Journal of Acquired Immune Deficiency Syndromes*, 3, 1990, pp. 293–5.

13 The geography in confidentiality

A global pandemic, reaching all societies and cultures, raises in yet another context the question of a global ethic for our planetary home. A thoughtful, and perhaps the first philosophical response was:

K-A. Apel, "The problem of a macroethic of responsibility to the future in the crisis of technical civilization," *Man and World*, 20, 1987, pp. 3–40.

The AIDS mapping projects in Los Angeles appear as poster-sized publications, and may be available from:

W. Bowen, *et al.*, "AIDS in LA," *Occasional Publications in Geography*, No. 4, California State University, Northridge, CA, 1989;

W. Bowen, B. Mlademich, "AIDS in LA 1983–89," *Occasional Publications in Geography*, No. 6, California State University, Northridge, CA, 1989.

A remarkable, empirically-rooted demonstration of the confidentiality question in its geographic coordinates is given in:

S. Openshaw, *et al.*, "An empirical study of the confidentiality crisis in the release of micro census data," Centre for Urban and Development Studies (CURDS), Newcastle University, UK.

Conventional time modeling estimates by statistical and differential models for England and Wales are given in:

Short-term Prediction of HIV Infection and AIDS in England and Wales, Department of Health, Welsh Office, Her Majesty's Stationary Office, London, 1988, p. 29,

while spatial diffusion approaches for Finland are undertaken in:

M. Löytönen, "The spatial diffusion of human immunodeficiency virus type 1 in Finland, 1982–1997," *Annals of the Association of American Geographers*, 81, 1991, pp. 129–51.

14 Education and planning: predicting the next maps

Approaches to spatio-temporal prediction open up a technical literature whose methods often require substantial, not to say super, computing facilities. A general introduction is available in:

J. Jones, E. Casetti, *Applications of the Expansion Method*, Routledge, London and New York, 1992;

S. Foster, "The expansion method: implications for geographic research,"

Professional Geographer, 43, 1991, pp. 131–42.

A reasonably accessible introduction to spatial adaptive filtering and AIDS diffusion is:

P. Gould, *et al.*, "AIDS: predicting the next map," *Interfaces*, 21, 1991, pp. 80–92;

with an alternative application by pioneers of the method in:

S. Foster, W. Gorr, "An adaptive filter for estimating spatially-varying parameters: application to modeling police hours spent in response to calls for service," *Management Science*, 32, 1986, pp. 878–89.

An introduction in Spanish is:

P. Gould, J. Kabel, "La epidemia de SIDA desde una perspectiva geografica," *GeoCritica*, 89, 1991,

while the educational purposes and possibilities are brought out in:

P. Gould, "Modeling the AIDS epidemic for educational intervention," in R. Ulack, W. Skinner (eds.), *AIDS and the Social Sciences: Common Threads*, University of Kentucky Press, Lexington, KY, 1991, pp. 30–44,

and in French in:

P. Gould, *et al.*, "Le SIDA: la carte animée comme rhétorique cartographique appliquée," *MappeMonde*, 1, 1990, pp. 21–6.

Animated maps showing the diffusion of AIDS in Ohio and Pennsylvania are:

P. Gould, *et al.*, *The Diffusion of AIDS*, videotape made in conjunction with WPSX-TV and the Deasy GeoCartographics Laboratory, University Park, PA, 1989,

while a television tape is also available for the entire United States:

P. Gould, *et al.*, *The Diffusion of AIDS in the United States, 1982–1990*, an animated cartographic presentation for television, University Park, PA 1992.

15 Herd immunity: riding the coat-tails of the HIV

The major publication raising this important question is:

W. O'Connor, "Herd immunity and the HIV epidemic," *Preventive Medicine*, 20, 1991, pp. 329–42,

but the documenting bibliography contains over 100 items in medical and epidemiological journals.

A related question, of great importance for the future development of the epidemic, is the way in which virulence may be related to forms of transmission. This relationship is discussed in:

P. Ewald, "Cultured vectors, virulence, and the emergence of evolutionary epidemiology", *Oxford Surveys in Evolutionary Biology*, 5, 1988, pp. 215–45;

P. Ewald, "Transmission modes and the evolution of virulence with special reference to cholera, influenza and AIDS," *Human Nature*, 2, 1991, pp. 1–30;

raising the further question of the importance of "external controls," the only controls we have until a cure or vaccine is discovered.

Index